Biomechanics of Soft Tissues

Biomechanics of Soft Tissues

Principles and Applications

Edited by
Adil Al-Mayah

CRC Press
Taylor & Francis Group
Boca Raton London New York

CRC Press is an imprint of the
Taylor & Francis Group, an **informa** business

CRC Press
Taylor & Francis Group
6000 Broken Sound Parkway NW, Suite 300
Boca Raton, FL 33487-2742

© 2018 by Taylor & Francis Group, LLC
CRC Press is an imprint of Taylor & Francis Group, an Informa business

No claim to original U.S. Government works

Printed on acid-free paper

International Standard Book Number-13: 978-1-4987-6622-7 (Hardback)

Visit the Taylor & Francis Web site at
http://www.taylorandfrancis.com

and the CRC Press Web site at
http://www.crcpress.com

Contents

Preface

The emerging paradigm of incorporating images and biomechanical properties of soft tissues has a proven potential as an integral part of the advancement of several medical applications, including image-guided radiotherapy and surgery, brachytherapy, and diagnostics. The subject of *Biomechanics of Soft Tissues* has been addressed in a number of journal papers and conferences, in addition to *anatomical site-specific* books. However, the urgent need for better understanding of the subject requires references that cover a wide scope of mechanical principles, properties, and applications.

This book starts by introducing the basics of soft tissue structures and the fundamentals of biomechanics using linear elastic, hyperelastic, viscoelastic, and poroelastic-modeling approaches (Chapter 1). To provide a quantitative sense of modeling parameters, this chapter also includes a list of mechanical parameters of human tissues. This chapter is followed by presenting different testing methods to measure these parameters (Chapter 2) where widely used direct mechanical tensile loading and indentation-testing methods are presented. The analytical procedures of these techniques are illustrated, in addition to application examples of soft tissue investigations. Chapter 3 presents the technique of elastography as an image-based method to measure mechanical properties of tissues. This chapter paves the way for some medical applications of soft tissue biomechanics. Chapter 4 discusses the role of biomechanical forces that are applied to cancer cells on cancer progression, tumor development, and consequently treatment. Specific attention was paid for biomechanics role on the therapeutic strategies in controlling cancer progression and the effects of some of the common chemotherapy drugs on biomechanics of cancer. Chapter 5 presents a unique integration of soft tissue biomechanics and imaging to address the challenging task of deformable image registration (DIR). It starts with a brief description of biomechanical-based DIR techniques developed in several anatomical sites. This is followed by an illustration of the role of biomechanical DIR in reducing geometric and dosimetric uncertainties in radiotherapy, improving the design of treatment, and augmenting our understanding of response to several clinical scenarios. Chapter 6 focuses on ultrasound-guided soft tissue interventions where a number of state-of-the-art technologies are presented. This chapter provides an excellent view of the potential ideas to incorporate soft tissue biomechanics to address a number of challenges associated with soft tissue deformation.

The combined medical and engineering expertise of contributors makes this book an excellent source of information and ideas for both engineering and medical professionals as challenges and expanding knowledge of different disciplines necessitate the adoption of the "*diplomacy of knowledge*" to expand collaboration and share ideas. This book provides medical professionals an insight into a wealth of modeling approaches, testing techniques, and mechanical characteristics that are frequently used by engineers. On the other hand, the presented medical applications provide engineers with a *glimpse* of amazing medical practices and encourage them to expand their roles in the medical field.

Editor

Professor Adil Al-Mayah is a member of the engineering school at the University of Waterloo, Waterloo, Ontario, Canada. He has done cutting-edge research integrating mechanics into imaging to accurately localize cancer tumors for radiotherapy applications. This technique has been successfully applied to different anatomical systems, including lungs, liver, head-and-neck, breast, and prostrate. He has an excellent publication record in top tier journals, refereed conferences, abstracts and presentations, invited talks, book chapters, and patents.

Contributors

Adil Al-Mayah
Department of Civil and Environmental
  Engineering
Mechanical and Mechatronics
  Engineering (Cross appointment)
University of Waterloo
Waterloo, Ontario, Canada

Kristy K. Brock
Department of Imaging Physics
The University of Texas M.D. Anderson
  Cancer Center
Houston, Texas

Aaron Fenster
Robarts Research Institute
The University of Western Ontario
and
Biomedical Engineering Graduate
  Program
The University of Western Ontario
and
Department of Medical Biophysics
The University of Western Ontario
London, Ontario, Canada

Derek Gillies
Robarts Research Institute
The University of Western Ontario
and
Department of Medical Biophysics
The University of Western Ontario
London, Ontario, Canada

Mohammad Kohandel
Department of Applied Mathematics
University of Waterloo
Waterloo, Ontario, Canada

Deirdre M. McGrath
NIHR Nottingham Biomedical Research
  Centre
Radiological Sciences
Queens Medical Centre
Nottingham, NG7 2UH, United Kingdom

Justin Michael
Robarts Research Institute
The University of Western Ontario
and
Biomedical Engineering Graduate
  Program
The University of Western Ontario
London, Ontario, Canada

Wanis Nafo
Department of Civil and Environmental
    Engineering
University of Waterloo
Waterloo, Ontario, Canada

Homeyra Pourmohammadali
Department of Applied Mathematics
University of Waterloo
Waterloo, Ontario, Canada

Jessica Rodgers
Robarts Research Institute
The University of Western Ontario
and
Biomedical Engineering Graduate
    Program
The University of Western Ontario
London, Ontario, Canada

Sivabal Sivaloganathan
Department of Applied Mathematics
University of Waterloo
Waterloo, Ontario, Canada

and

Center for Mathematical Medicine
Fields Institute for Research in
    Mathematical Sciences
Toronto, Ontario, Canada

Michael Velec
Techna Institute and Princess Margaret
    Cancer Centre
University Health Network
Toronto, Canada

1

Mechanical Characteristics of Soft Tissues

Adil Al-Mayah

1.1 Introduction

1.1.1 General

Soft tissues are defined as the tissues that support and connect body structures. They include skin, muscles, fat, tendons, ligaments, blood vessels, nerves, cartilages, and other tissue matrices. In some cases, they are simply defined as body tissues that exclude hard tissues such as bones, teeth, and nails. As bones, a major component of nonsoft tissues, represent 12%–15% of the human body mass, it can be concluded that most of the human body is composed of soft tissues. Soft tissues are known for high flexibility and soft mechanical properties, differentiating them from mineralized stiff tissues, such as bones (Holzapfel 2001).

1.1.2 Source of Mechanical Response in Soft Tissues

Modeling biological phenomena and material behavior can be performed at the atomic, molecular, microscopic, and macroscopic scales. The mechanical response of tissues can be well addressed at the macroscopic scale of biological modeling and to a limited extent at the microscopic (multicellular) scale. Therefore, these scales will be the focus of this section.

At the cellular level, each cell consists of a cellular membrane, cytoplasm, and nucleus. The membrane and structural cytoplasmic component, known as the cytoskeleton, are the main contributors to the structural performance of cells. The membrane separates the intra- and extracellular environments and plays a role in the interaction between these two environments. The cytoskeleton provides structural integrity of cells. It consists of three types of filaments: (1) actin (8 nm diameter), (2) an intermediate rope-like structure (10 nm diameter), and (3) microtubules (25 nm diameter). Actin filaments are stiffer in extension than microtubules but they rupture at a much lower extension. The intermediate filaments exhibit an intermediate extensional stiffness at lower extensions, but they can sustain much larger extensions than the other two types of filaments while exhibiting a nonlinearly stiffening response. The microtubules are long cylinders that exhibit high bending stiffness compared to other filaments.

At the tissue level, typical tissues consist of three main components: (1) epithelial, (2) stromal, and (3) mesenchymal cells. Epithelium is one of the four basic animal tissues along with connective, muscle, and nervous tissues. It is composed of packed epithelial cells arranged in varying numbers of layers that line the body cavities and surfaces and form glands. Epithelial tissues' main functions include protection and secretion. The epithelial cells are attached to each other at many locations through adherence junctions, tight junctions, and spot-like adhesion (desmosomes). These epithelial cells rest on a membrane through a keratin-based cytoskeleton and adhesion-based junctions (hemidesmosomes). The thin semipermeable membrane separates the epithelium from the stroma. The stroma is a loose connective tissue that may rest on layers of muscles or bones. It is composed of extracellular matrix (ECM), blood vessels, nerves, and lymphatic vessels. The ECM consists of a scaffolding of fibers, such as collagen and elastin, embedded in a mixture of water and glycoproteins, and is of particular interest in terms of mechanical performance.

Collagen represents the main structural component of hard and soft tissues in animals, and is responsible for the mechanical performance and strength of many elements of the human body, including blood vessels, tendons, and bones (Fung 1993). Interestingly, Fung (1993) compared its role to the role of steel in the advancement of many aspects of our civilization. Therefore, the performance of many soft tissues can be attributed to their collagen fiber content. In addition, the high water content of soft tissues plays a vital role in their mechanical behavior due to its incompressibility and viscous nature, which contribute to viscoelastic and poroelastic performance.

1.1.3 General Mechanical Behavior

Soft tissues have a complex structure that is generally described as a "nonlinear, inelastic, heterogeneous, anisotropic character that varies from point to point, from time to time and from individual to individual" (Humphrey 2003). For example, the generalized stress–strain curve of a soft tissue under a simple stretching load with a constant loading rate exhibits three main regions before rupture, depending on the intensity of the applied load. Generally, it shows a linear behavior under low loading with low stiffness, which becomes nonlinear as load increases. As the applied load increases further, a linear behavior with high stiffness is observed.

Changing the rate of loading can alter this behavior, reflecting the time-dependent characteristics of soft tissues. Therefore, different mechanical models have been developed to capture the behavior of soft tissues under loading. These models include elastic (linear), hyperelastic (nonlinear elasticity), viscoelastic (time-dependent), and poroelastic (biphasic) types. Each model has its own applications, spanning from assessment of organ's deformation to drug distribution inside the tissues. The details of the required outcomes of these models may necessitate sophisticated material properties. For example, for soft tissues experiencing little deformation during regular physiological activities, linear elastic properties are sufficient. However, hyperelastic properties are recommended for accurate estimation of organ deformation when large deformation is expected. In addition, viscoelastic and poroelastic properties are required for time-dependent and multiphasic tissue responses.

Furthermore, materials can generally be characterized according to their homogeneity (location-dependent properties) and isotropy (direction-dependent properties). To provide a better understanding of isotropy and homogeneity, these concepts will be discussed using a linear elasticity model as an illustrative example.

1.2 Linear Elasticity

1.2.1 Model Description

Linear elastic models have been widely used to characterize the behavior of soft tissues, as they have for many other tissues such as metals. They may not provide a full representation of material behavior because it is limited to a low-load level. However, their popularity in mechanical modeling of soft tissues is due to their simplicity. In addition, these models are sufficient for a number of modeling aspects of organ deformation, as in the case of deformable image registration for image-guided interventions where regular physiological deformation of some organs is small. However, it is not a favored model in applications where large deformation is expected, such as traumatic automotive and sport incidents.

The applied load and the corresponding deformation data can be established using tensile, compressive, or indentation-testing methods. For a sample with a cross-sectional area of A and length L and subjected to a uniaxial tensile loading of P, the force is often translated into a stress (σ), which is the ratio of the applied load to the cross-sectional area (i.e., $\sigma = P/A$). In addition, the deformation (ΔL) is

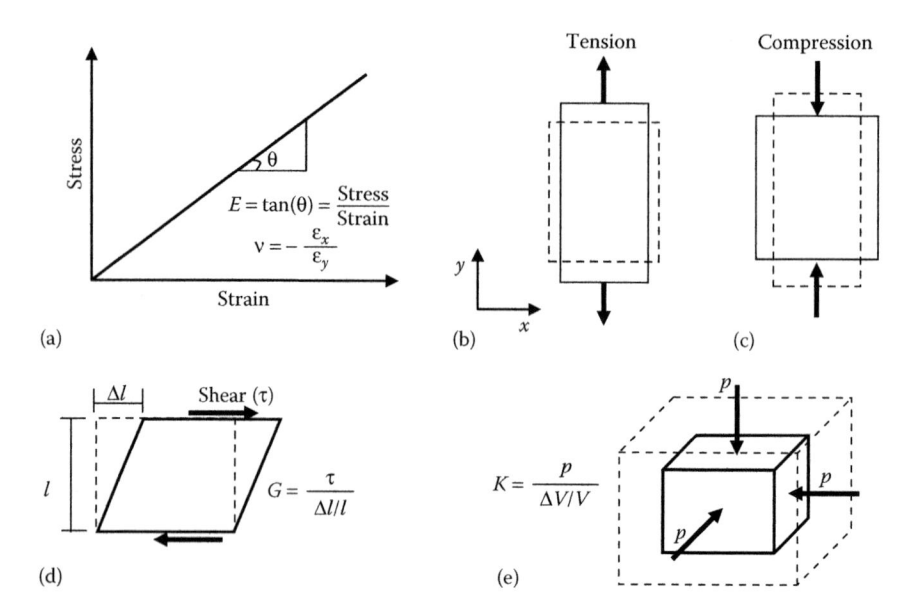

FIGURE 1.1 (a) Stress–strain plot of elastic material under (b) tensile and (c) compression loads where the tangent of the plot represents the modulus of elasticity, and the negative ratio of lateral to longitudinal strains is the Poisson's ratio, (d) rotational deformation caused by shear, and (e) volumetric deformation under equal pressure loading. (With kind permission from Taylor & Francis: *Imaging in Medical Diagnosis and Therapy* 2013, 85–94, Al-Mayah, A. and Brock, K.)

represented by a strain (ε), which is the ratio of the length change to the original length (i.e., $\varepsilon = \Delta L/L$), as illustrated in Figure 1.1a–c.

The slope of the stress–strain curve is called the elastic modulus ($E = \sigma/\varepsilon$), often known as Young's modulus (Figure 1.1a). In addition, the compressibility factor represented by Poisson's ratio (v) is the second parameter required to describe the material's behavior that can be calculated as the negative ratio of the transverse strain (ε_x) to the longitudinal strain (ε_y) in the direction of the applied load ($v = -\varepsilon_x/\varepsilon_y$). In some cases, mechanical properties of tissues are reported in terms of shear modulus (G), which represents the ratio of the shear stress (τ) to the shear strain represented by the angular deformation of the distorted shape ($\approx \Delta l/l$), as shown in Figure 1.1d. The shear modulus can be written in terms of the elastic modulus and Poisson's ratio, where $G = 0.5E/(1 + v)$. In other cases, a material experiences volumetric changes (ΔV) because of equal pressures applied from all directions, as shown in Figure 1.1e. The bulk modulus ($K = p/(\Delta V/V)$) is used in this case, where ($K = E/[3(1-2v)]$).

1.2.2 Elastic Properties of Human Tissues

Linear elastic properties of some human soft tissues are listed in Table 1.1. Although most of the soft tissue properties reported in the literature are *ex vivo*, and are often

TABLE 1.1 Elastic Modulus of Different Human Soft Tissues

Tissue	Testing Method	Average Elastic Modulus (kPa)	Reference
Brain	MRE	12.9 kPa white matter 15.2 kPa gray matter	Uffmann et al. 2004
Arteries (coronary)	Uniaxial tensile (*ex vivo*)	1480–1550 kPa healthy 3770–4530 kPa atherosclerotic	Karimi et al. 2013
Ascending thoracic aorta	Uniaxial tensile (*ex vivo*)	2982 kPa healthy 7641 kPa Marfan syndrome	Jarrahi et al. 2016
Breast	MRE	17.1–23.5 kPa fat 24.2–30.3 kPa fibroglandular	Van Houten et al. 2003
Esophagus	Ultrasonography and manometry	4.9–13.6 kPa	Takeda et al. 2002
Liver	USE	0.64–1.08 kPa (liver) 3–12.1 kPa (tumor) 1.11–4.93 kPa (fibrosis)	Yeh et al. 2002
	Indentation	270 kPa	Carter et al. 2001
	Aspiration	6.0 kPa	Muller et al. 2009
	Aspiration	20 kPa (long term) 60 kPa (instantaneous)	Nava et al. 2008
Liver with fibrosis	Transient elastography	7.2–18.2 kPa	Marcellin et al. 2009
	Elastography	2.8–16.5 kPa (stage 0–2) 6.8–69.1 kPa (stage 3,4)	Corpechot et al. 2006
Prostate with PBH	Sonoelastography	24.1 kPa peripheral 32.2 kPa transitional	Zhang et al. 2014
Parotid gland	USE	26 kPa (healthy) 146.6 kPa (malignant) 88.7 kPa (benign)	Wierzbicka et al. 2013
Thyroid cancer	Compression	45 kPa	Lyshchik et al. 2005

Note: MRE = magnetic resonance elastography, USE = ultrasonic elastography.

animal tissue properties, efforts have been made to report human and *in vivo* tissue properties. Different testing methods were used, including direct mechanical tensile and indentation tests, in addition to image-based elastography. More details on these methods will be presented in the upcoming chapters 2 and 3. The investigated parameters include the elastic modulus (E) or shear modulus (G), and Poisson's ratio (ν); however, the elastic modulus is one of the most widely reported measurements. Although, soft tissues are mostly incompressible or nearly incompressible ($\nu \leq 0.5$), some investigations have reported Poisson's ratio values. For example, Lai-Fook and Hyatt (2000) experimentally measured Poisson's ratio for lung parenchyma, in addition to the shear modulus. They found that Poisson's ratio was age related and increased from 0.41 to 0.45 as age increased. In addition, the effective shear modulus of human lungs (*in vivo*), measured using MR elastography, was affected by the volume of inflation ($G = 3.45$ kPa at residual volume, and 10.75 kPa at the total lung capacity) (Mariappan et al. 2011).

Significant variations are observed among the reported data, even when looking at the same organs. However, this is expected with the range of different individuals, testing methods and procedures, and load/strain ranges applied across studies. For example,

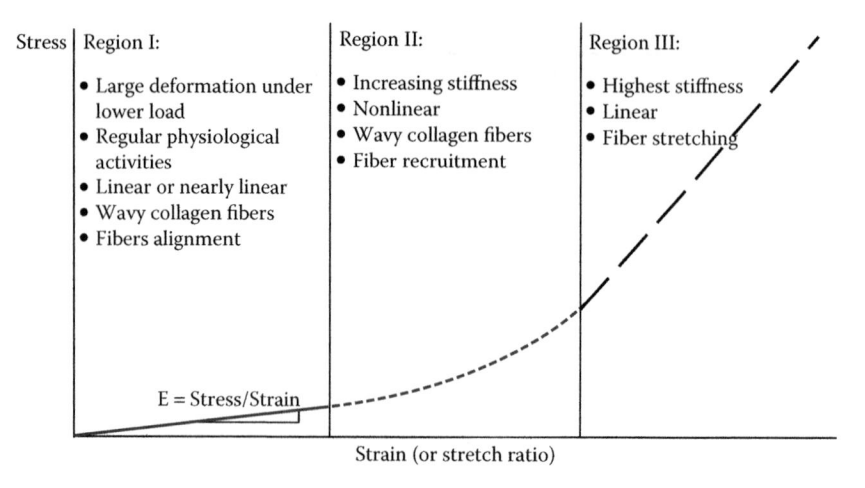

FIGURE 1.2   General nonlinear behavior of soft tissues stretched under a constant loading rate.

the modulus of elasticity is load/strain dependent, as shown in Figure 1.2; however, few details have been provided in some papers.

1.3 Hyperelasticity

1.3.1 Model Description

Many materials perform nonlinearly under loading, as characterized by the nonlinear stress–strain (or stress–stretch ratio) relationship. Typical nonlinear stress–strain relationships of hyperelastic material combine three regions, as shown in Figure 1.2. In the first region (Region I), the tissue experiences large deformation under relatively low loading (low stiffness) in a linear or nearly linear pattern attributed to the removal of waves of collagen fibers in relaxed tissues. Most of the typical physiological activities are within this region. This is followed by a nonlinear region (Region II) with an increasing stiffness due to the initial recruitment of stiff collagen fibers. As the load further increases (Region III), the tissue exhibits a stiffer behavior that is mainly characterized by a linear stress–strain relationship as stiff fibers are stretched and actively participate in carrying the applied load.

Different approaches have been proposed to capture this nonlinear material performance, reaching back to 1847, when Wertheim showed a nonlinear stress–strain relationship of animal tissues that deviated from the linear elastic Hooke's law. This was formulated by direct nonlinear stress–strain equations or through the use of strain-energy functions.

More direct nonlinear equations have been proposed (Fung 1993), as listed in Table 1.2; however, they were not intended to represent three-dimensional (3D) stress states (Fung 1993). Regardless of their long history, these equations are not widely used in soft-tissue characterization.

On other hand, strain-energy or potential-energy functions (W) are widely used to capture a wide range of elastic finite deformation. These models are applied to rubber

TABLE 1.2 Nonlinear Stress–Strain Energy Function
Used to Model Biological Tissues

Reference	Model
Wertheim (1847)	$\varepsilon^2 = a\sigma^2 + b\sigma$
Morgan (1960)	$\varepsilon = a\sigma^n$
Kendedi et al. (1964)	$\sigma = k\varepsilon^d, \sigma = B(e^{m\varepsilon} - 1)$
Ridge and Wright (1964)	$\varepsilon = C + k\sigma^b, \varepsilon = x + y\log\sigma$
Hoeltzel et al. (1992)	$\sigma = \alpha(\varepsilon - \varepsilon_s)^\beta$

and rubber-like materials, also known as hyperelastic materials or Green elastic material (named after Green in 1839), where they are characterized as incompressible or nearly incompressible materials (i.e., Poisson's ratio $\nu \approx 0.5$). This works well for most soft tissues because of their incompressibility nature associated with their high water content.

The strain-energy function (W) represents a measure of energy stored in a material due to the applied strain. The relationship between the strain-energy function and the material deformation is represented by the stretch ratios in principle directions (λ_1, λ_2, and λ_3). Three strain invariants (I_1, I_2, and I_3) are used to represent the stretches, as shown in Equation 1.1. These strain invariants are the same regardless of the applied coordinate system.

$$I_1 = \lambda_1^2 + \lambda_2^2 + \lambda_3^2$$

$$I_2 = \lambda_1^2\lambda_2^2 + \lambda_2^2\lambda_3^2 + \lambda_3^2\lambda_1^2 \tag{1.1}$$

$$I_3 = \lambda_1^2\lambda_2^2\lambda_3^2$$

For incompressible materials, the value of the third invariant (I_3) is 1.

Different energy functions have been developed to capture the performance of hyperelastic materials. Given a wide variation of soft tissue responses to mechanical loading, the efficiency of each of these energy functions depends on its ability to capture the full behavior of the material with a minimum number of parameters. In other words, if the proposed model results fit with the experimental data of a material subjected to a specific loading, the model is considered efficient to predict the material behavior under this specified loading condition. Reviews on some of these models can be found in Martins et al. (2006) and Boyce and Arruda (2000). Some of the popular hyperelastic models used in biomechanical modeling are briefly described below:

1. *Mooney–Rivlin model* is an early model that was used to capture the nonlinear behavior of rubber-like materials (Martins et al. 2006). It is known for its high accuracy, especially for cases in the median range of strain (200%–250%). The strain energy function is

$$W = C_{10}(I_1 - 3) + C_{01}(I_2 - 3) \tag{1.2}$$

where C_{10} and C_{01} are material constants in stress units (e.g., N/mm²).

2. *Polynomial model* was proposed by Rivlin as an extension of the Mooney–Rivlin model. The model is proposed in a form of polynomial series as follows:

$$W = \sum_{i,j=0}^{\infty} C_{ij} (I_1 - 3)^i (I_2 - 3)^j \tag{1.3}$$

where C_{ij} is a material parameter.

3. *Neo–Hookean model* considers only the first term of the Rivlin model. It is suitable for modeling small strains of 150%. It has been widely used in modeling biological tissues because of its accuracy in capturing material behavior under different loading (Marckmann and Verron 2006). In addition, it is recognized for its simplicity because only one parameter is required.

$$W = C_{10} (I_1 - 3) \tag{1.4}$$

4. *Yeoh model* uses higher order terms of I_1 or I_2 to account for a wider spectrum of deformation. Adding a higher order of I_1 was shown to accurately model large deformation loading cases. In addition, the effect of I_2 on the accuracy of material characteristics was also minimal. Therefore, the Yeoh model focuses on three orders of I_1 only, as illustrated in Equation 1.5 for incompressible material models:

$$W_{\text{Yeoh}} = C_{10} (I_1 - 3) + C_{20} (I_1 - 3)^2 + C_{30} (I_1 - 3)^3 \tag{1.5}$$

5. *Arruda and Boyce model* considers higher order terms for incompressible materials:

$$W = \mu \sum_{i=1}^{5} \frac{C_i}{\lambda^{2i-2}} \left(I_1^i - 3^i \right)^i \tag{1.6}$$

where $C_1 = 1/2$, $C_2 = 1/20$, $C_3 = 11/1050$, $C_4 = 19/7000$, $C_5 = 519/673750$, λ is the locking stretch ratio (unitless), and μ is known as the initial shear modulus.

6. *Ogden model* uses principal stretches in the strain-energy function instead of the strain invariants used in other models (Ogden 1984). This model has been used to model large deformation cases:

$$W_o = \sum_{i=1}^{n} \frac{\mu_i}{\alpha_i} \left(\lambda_1^{\alpha_i} + \lambda_2^{\alpha_i} + \lambda_3^{\alpha_i} - 3 \right) \tag{1.7}$$

where μ_i (stress units) and α_i (unitless) are real numbers representing material parameters, whereas (n) is a positive integer.

7. *Veronda–Westmann* model has been used to model incompressible materials:

$$W_{\text{VW}} = C_1 \left(e^{\alpha(I_1 - 3)} - 1 \right) - C_2 (I_2 - 3) \tag{1.8}$$

where:

C_1 and C_2 are material constants with stress units

α is a unitless parameter

8. *Fung model* (Fung 1975) is an exponential pseudostrain energy function $(\rho_0 W)$, described as follows:

$$\rho_0 W = \frac{C}{2}\left[e^{\left(a_1\varepsilon_x^2 + a_2\varepsilon_y^2 + 2a_4\varepsilon_x\varepsilon_y\right)} + e^{\left(a_1\varepsilon_x^2 + a_2\varepsilon_z^2 + 2a_4\varepsilon_x\varepsilon_z\right)} + e^{\left(a_1\varepsilon_z^2 + a_2\varepsilon_y^2 + 2a_4\varepsilon_z\varepsilon_y\right)} \right] \tag{1.9}$$

where:

ε_x, ε_y, and ε_z are strain components

c, a_1, a_2, and a_4 are material constants

Although this model was introduced in the biomechanics literature several decades ago, its implementation into finite element (FE) simulations has been limited. The key barriers to numerical implementations are inherent numerical instability and convergence (Sun et al. 2005).

Although most hyperelastic models are isotropic, some anisotropic models with different degrees of anisotropy have been introduced, including transversely isotropic (Humphrey et al. 1990a, b; Weiss et al. 1996; Gardiner and Weiss 2001) and orthotropic (Tong and Fung 1976; Chew et al. 1986; Fung 1993; Criscione et al. 2003) models. However, the numerical convergence problem is a barrier for using these in numerical modeling (Sun et al. 2005).

1.3.2 Hyperelastic Properties of Human Tissues

A nonlinear stress–strain relationship has been reported for different organ tissues. Zeng et al. (1987) experimentally investigated the mechanical properties of excised human lung tissues using biaxial loading. A nonlinear hyperelastic model was applied to characterize the nonlinear stress–strain relationship. Similar behavior was observed by Budday et al. (2017) in their extensive investigation of *ex vivo* human brain tissues, where shear, tension, and compression tests were conducted to investigate hyperelastic properties of different brain regions. The experimental data fit well with the modified one-term Ogden strain-energy function by comparison with other energy functions such as Neo–Hookean, Mooney–Rivlin, Demiray, and Gent. The properties varied among different brain regions and testing methods as reported in Table 1.3.

Similar variations were also reported among different regions of the breast, as reported by Samani and Plewes (2004), and among different types of breast cancers (O'Hagan and Samani 2009), including ductal carcinoma *in situ*, invasive mucous carcinoma, invasive lobular carcinoma, and low-, medium-, and high-grade invasive ductal carcinomas. It was clearly shown that cancerous tissue parameters were larger than healthy tissues in some cases by as much as two orders of larger magnitude (O'Hagan and Samani 2009). Table 1.3 provides a sample of hyperelastic properties and models of human tissues.

TABLE 1.3 Human Tissue Hyperelastic Parameters

Hyperelastic Model and Parameters	Testing Method	Reference
Brain		
$W = C_{10}\left(I_1 - 3\right) + C_{30}\left(I_1 - 3\right)^3$ $C_{10} = 0.24$ kPa, $C_{30} = 3.42$ kPa	Aspiration test (*in vivo*)	Schiavone et al. 2009
$\psi_{\text{Ogden}} = 2\mu/\alpha^2\left(\lambda_1^\alpha + \lambda_2^\alpha + \lambda_3^\alpha - 3\right)$ Parameter range represents different regions of brain: $\mu = 0.33$ to 1.06 kPa, $\alpha = -22.0$ to -24.6 (shear test) $\mu = 0.33$ to 1.16 kPa, $\alpha = -25.6$ to -38.9 (tension test) $\mu = 0.47$ to 1.63 kPa, $\alpha = -11.4$ to -16.5 (compression test)	Tension, compression, and shear (*ex vivo*)	Budday et al. 2017
Breast		
$W = \sum_{i+j=1}^{N} C_{ij}\left(I_1 - 3\right)^i\left(I_2 - 3\right)^j$ ($C_{10} = 0.31$, $C_{01} = 0.30$, $C_{11} = 2.25$, $C_{20} = 3.80$, $C_{02} = 4.72$) kPa (adipose) ($C_{10} = 0.33$, $C_{01} = 0.28$, $C_{11} = 4.49$, $C_{20} = 7.72$, $C_{02} = 9.45$) $\times$ 10^{-4} kPa (fibroglandular)	Indentation + Inverse FEM	Samani and Plewes 2004
Lungs		
$\rho_0 W = \dfrac{C}{2}\left[e^{\left(a_1\varepsilon_x^2 + a_2\varepsilon_y^2 + 2a_4\varepsilon_x\varepsilon_y\right)} + e^{\left(a_1\varepsilon_x^2 + a_2\varepsilon_z^2 + 2a_4\varepsilon_x\varepsilon_z\right)} + e^{\left(a_1\varepsilon_z^2 + a_2\varepsilon_y^2 + 2a_4\varepsilon_z\varepsilon_y\right)}\right]$ $c = 11.8$ g/cm, $a_1 = 0.43$, $a_2 = 0.56$, $a_4 = 0.32$	Biaxial tensile (*ex vivo*)	Zeng et al. 1987
$\rho_0 W = \dfrac{C}{2\Delta} e^{\left(\varepsilon I_1^2 + \beta I_2\right)}$ $\dfrac{C}{\Delta} = 3.06 \pm 0.84 K. \dfrac{\text{dyn}}{\text{cm}^2}$, $\alpha = 4.47 \pm 1.94$, $\beta = -4.2 \pm 2.55$		Gao et al. 2006
Liver		
$W = C_{10}\left(I_3^{-\frac{2}{3}} I_1 - 3\right) + C_{20}\left(I_3^{-\frac{2}{3}} I_1 - 3\right)^2 + \dfrac{K_0}{2}\left(I_3 - 1\right)^2$ $C_{10} = 9.85$ kPa, $C_{20} = 26.29$ kPa, $K_0 = 10^4$ kPa	Aspiration test + Inverse FEM (*in vivo*)	Nava et al. 2008
Liver capsule $W = C_{20}\left(I_1 - 3\right)^2$ one-term polynomial model $C_{20} = 114 \pm 40$ kPa $W = a\left[e^{b\left(I_1 - 3\right)} - 1\right]$ Exponential model (Demiray) $a = 71 \pm 29$ kPa, $b = 1.8 \pm 0.5$ kPa $I_1 = \lambda_1^2 + \lambda_2^2 + \dfrac{1}{\lambda_1^2\lambda_2^2}$	Inflation test	Brunon et al. 2011

1.4 Viscoelasticity

1.4.1 Model Description

In previous linear elastic and hyperelastic modeling sections, the time element was not considered as a contributing factor to the mechanical behavior of soft tissues. However, duration and rate of loading (force/unit time) affect the mechanical behavior of soft tissues given their high-fluid (viscous) content. Depending on the application of the mechanical properties of soft tissues, there is a debate on the importance of this factor. However, most of this debate focuses on the significance of including the time factor and not on the intrinsic characteristic of tissues.

Figure 1.3 illustrates the time-dependent response of pure elastic, pure viscous, and viscoelastic materials, which combines pure elastic and pure viscous responses. The differences in response can be recognized in the loading and unloading stages. The loading and unloading response of elastic material is independent of time. This is clearly demonstrated by the immediate response of these materials to both loading and unloading conditions. During loading, the load is instantaneously transferred to the material. Similarly, the material responds to unloading immediately after the load is removed. However, viscous material responses at the loading stage are a function of time. In this case, time is needed to transfer the full load to the material, often referred to as a velocity of deformation. In addition, the material is deformed permanently, even after the load is removed during unloading stage. On the other hand, viscoelastic material responses are a mixture of pure elastic and viscous responses, where deformation is a function of time during both the loading and unloading conditions. However, unlike pure viscous materials, the strain drops suddenly after the load removal in the unloading stage, but requires time to fully recover and return to its original configuration. Given their pattern of loading response, viscoelastic materials are modeled using spring and dashpot to model the elastic and viscoelastic response, respectively.

The time-dependent response of viscoelastic materials can be divided into different types: creep, relaxation, strain-rate, and hysteresis, as shown in Figure 1.4. In creep, under a constant load, the material continues to experience deformation (strain) over time. On the other hand, the material relaxes when it is subjected to a constant deformation as stress drops. In addition, the response of the material to loading is dependent

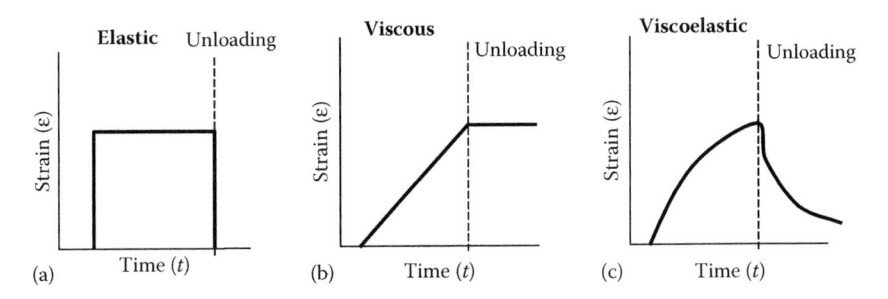

FIGURE 1.3 Time-dependent response of (a) elastic, (b) viscous, and (c) viscoelastic materials.

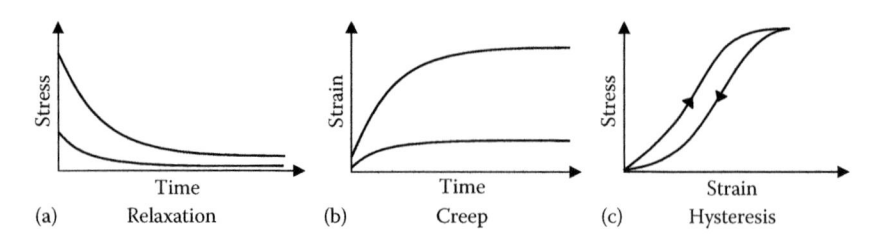

FIGURE 1.4   (a) Stress–time behavior of a viscoelastic material under two constant strain levels, (b) strain–time behavior of a viscoelastic material subjected to two constant stress levels, and (c) Hysteresis of stress–strain plot of a viscoelastic material under cyclic loading. (With kind permission from Taylor & Francis: *Imaging in Medical Diagnosis and Therapy* 2013, 85–94, Al-Mayah, A. and Brock, K.)

on the duration of the applied load as it is capable of carrying higher loads under faster strain application (i.e., higher strain rate). Therefore, the strain/loading rate is often reported with viscoelastic material properties. In hysteresis, the viscoelastic material dissipates energy when it is subjected to loading–unloading cycles where the loading path is different from that of unloading.

There are three types of viscoelasticity: (1) linear, (2) quasi-linear, and (3) nonlinear. Linear viscoelasticity is used in a wide range of applications due to its simplicity; hence, it is the focus of this section. Linear viscoelastic models generally include a solid-related characteristic (e.g., spring), in addition to the fluid component (e.g., damper or dashpot). Different arrangements and numbers of these components have been proposed to create a number of viscoelastic models. Some of these common models are presented here, including Maxwell, Kelvin–Voigt, and standard linear solid (Zener model), as listed in Table 1.4.

The Maxwell model is the simplest model, where it consists of a spring and a dashpot arranged alongside each other. Therefore, both spring and dashpot are subjected to the same load. It accurately predicts the relaxation response, but not the creep response, as described by

$$\dot{\varepsilon} = \dot{\varepsilon}_s + \dot{\varepsilon}_d = \frac{1}{E}\frac{d\sigma}{dt} + \frac{\sigma}{\eta} \tag{1.10}$$

where:

$\dot{\varepsilon}_s = \dfrac{d\varepsilon_s}{dt}$, $\dot{\varepsilon}_d = \dfrac{d\varepsilon_d}{dt}$, and $\dot{\varepsilon} = \dfrac{d\varepsilon}{dt}$

ε_s and ε_d are spring and dashpot strains, respectively, produced by the applied stress (σ)

E and η are the spring and dashpot constants, respectively

In the Kelvin–Voigt model, both spring and dashpot are subjected to the same displacement due to their parallel arrangement. It is worth mentioning that the Kelvin–Voigt model shows a unique relaxation response to a rigid body in sudden loading because the dashpot does not move under sudden loading. It is well suited to the prediction of creep. The stress is calculated by

TABLE 1.4   Linear Viscoelastic Models

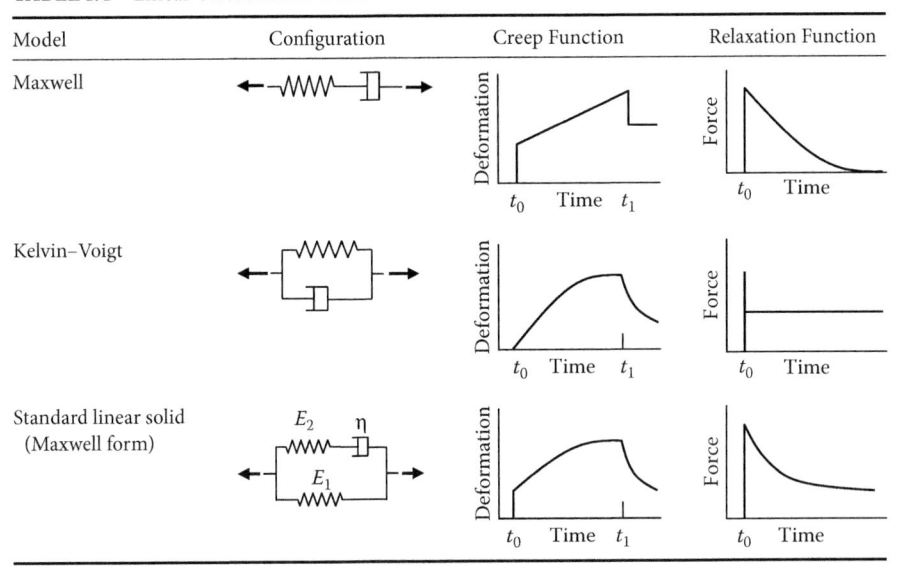

Model	Configuration	Creep Function	Relaxation Function
Maxwell			
Kelvin–Voigt			
Standard linear solid (Maxwell form)			

Source: Al-Mayah, A. and Brock, K. 2013. *Imaging in Medical Diagnosis and Therapy*, CRC Press, Taylor & Francis Group, Boca Raton, FL, 85–94.

$$\sigma = \sigma_s + \sigma_d = E\varepsilon_s + \eta \frac{d\varepsilon_d}{dt} \tag{1.11}$$

where:

ε_s and ε_d are spring and dashpot strains, respectively

E and η are the spring and dashpot constants, respectively

As for the standard linear solid model using the Maxwell form, the following relationship is applied:

$$\frac{d\varepsilon}{dt} = \frac{E_2}{\eta(E_1 + E_2)}\left(\frac{\eta}{E_2}\frac{d\sigma}{dt} + \sigma - E_1\varepsilon\right) \tag{1.12}$$

where E_1 and E_2 are the spring constants aligned and parallel to the dashpot, respectively, as shown in Table 1.4.

1.4.2 Viscoelastic Properties of Human Tissues

Different studies have applied viscoelastic properties to model soft tissues. The reported constants are normally the spring (E) and dashpot (μ) constants. However, in some cases, a shear modulus (Sinkus et al. 2005; Klatt et al. 2007) and a complex elastic modulus representing the frequency-dependent modulus of elasticity (Zhang et al. 2008) have also been reported.

A wide range of human tissues have been investigated to find viscoelastic properties. Among the most widely studied are brain tissues. This is mainly related to the fact that brain tissues experience a number of time-dependent large deformations, mostly associated with the fluid movement. Some of these deformations are slow, such as those associated with hydrocephalus, and convolutional development (Franceschini et al. 2006), in addition to surgery induced deformation. Other deformations are large and fast, such as traumatic injuries. Therefore, viscoelastic properties are required for time-dependent reactions. These properties are addressed in a simple linear viscoelastic form (Klatt et al. 2007; Green et al. 2008; Zhang et al. 2011) and poroviscoelastic form (Franceschini et al. 2006; Mehrabian and Abousleiman 2011). Similar investigations have been conducted on other human soft tissues, including the breast (Krouskop et al. 1998; Sinkus et al. 2005), liver (Klatt et al. 2007; Asbach et al. 2008), lung (Zhang et al. 2008), and prostate (Krouskop et al. 1998).

A sample of linear viscoelastic parameters of human tissues is listed in Table 1.5.

TABLE 1.5   Human Tissue Viscoelastic Parameters

Tissue	Testing Method	Parameters	Reference
Brain	MRE (*in vivo*)	$\mu_1 = 0.84$ kPa, $\mu_2 = 2.03$ kPa, $\eta = 6.7$ Pa s Zener_ μ_1, μ_2 shear modulus	Klatt et al. 2007
Breast + pathologies	MRE (*in vivo*)	$G = 0.87$ kPa, $\eta = 0.55$ Pa s Voigt model $G = 2.9$ kPa, $\eta = 2.4$ Pa s Cancer $G = 1.3$ kPa, $\eta = 2.1$ Pa s Fibroadenoma $G = 1.2$ kPa, $\eta = 0.8$ kPa Mastopathy	Sinkus et al. 2005
Liver	MRE (*in vivo*)	$\mu_1 = 1.36$ kPa, $\mu_2 = 1.86$ kPa, $\eta = 5.5$ Pa s Zener_ μ_1, μ_2 shear modulus	Klatt et al. 2007
Liver with fibrosis	MRE (*in vivo*)	$E_1 = 1.16$ kPa, $E_2 = 1.97$ kPa, $\eta = 7.3$ Pa s	Asbach et al. 2008
Lung	Tension (*ex vivo*)	$E = 6.8$ kPa, $R = 3.1$ kPa s, $\eta = 0.072$ with airways $E = 9.6$ kPa, $R = 4.0$ kPa s $\eta = 0.062$ without airways $E = 74$ kPa, $R = 3.5$ kPa s total (R = resistance, $\eta = 0.075$ hysteresivity tissue damping/elastance)	Dolhnikoff et al. 1998
Prostate	Sonoelastography + compression test (*in vivo* + histology)	$E^* = 15.9$ kPa, $\eta = 3.61$ kPa s$^\alpha$, $\alpha = 0.2154$ healthy $E^* = 40.4$ kPa, $\eta = 8.65$ kPa s$^\alpha$, $\alpha = 0.2247$ cancerous (Kelvin–Voigt model) $\sigma(t) = E_0\varepsilon(t) + \eta D^\alpha\big[\varepsilon(t)\big]$	Zhang et al. 2008

E^* (complex Young's modulus) is a frequency-dependent Young's modulus.

In the aforementioned elasticity, hyperelasticity, and viscoelasticity models, a material has been considered as *uniphase* solid and its response to external loads or deformation is modeled as a *lumped* relationship. Although this assumption is a sufficient representation to capture mechanical behavior of soft tissues, it does not fully cover some applications where the *biphases* model is needed as in the case of drug-distribution investigations. This biphasic model is based on the fact that soft tissues contain a high percentage of water, occupying a large percentage of its total volume in spaces that may be referred to as *pores*.

1.5 Poroelasticity

1.5.1 Model Description

The structure of soft tissues is generally considered biphasic, consisting of a porous solid phase and a fluid phase. Therefore, a poroelastic model has been used to model tissues. As the solid plays a major role in load-carrying capacity of the porous materials, different important functions are attributed to the fluid constituent in tissues. These functions include transport of nutrients from the vascular system to cells, removal of waste from cells, preventing friction in cartilage, and drug delivery and distribution, in addition to its role in load transfer. In general, poroelastic material behavior is similar to that of viscoelastic materials where it experiences creep under a constant stress and relaxation under constant strain steps.

Poroelastic theory was developed to model soil consolidation in 1923 and 1925 by Terzaghi (Terzaghi 1925), who assumed a one-dimensional (1D) consolidation case and incompressible solid and fluid constituents. The 1D assumption was adopted by assuming the soil as a laterally confined material, thus experiencing only uniaxial deformations. On the basis of the incompressibility assumption, deformation of soil under compression is mainly caused by the rearrangement of particles and not by the compression of solid particles and pore fluid. In other words, in saturated soil, the volume change of a material can occur only by the net flow of fluid out of pores. This assumption provides a good approximation for the behavior of highly incompressible soft soils such as clay and sands, where compressibility of stiff solid particles is negligible as compared to that of the whole porous material.

Biot (1941) further developed the poroelastic theory, presenting a 3D theory of linear elastic deformation of porous media, taking into account the compressibility of the constituents. In the year between 1955 and 1973, Biot proposed further developments by including an anisotropic case, dynamic response, and nonlinear elasticity. Verruijt (1969) extended the formulation of this theory to soil mechanics problems. In addition, Rice and Cleary (1976) reformulated the theory in terms of drained and undrained behavior, and fluid-filled porous materials.

Another approach was proposed to model poroelastic material using the theory of mixtures. This approach is based on diffusion models developed using fluid and thermodynamics principles (Truesdell and Toupin 1960; Bowen 1980), where each constituent has a number of specified characteristics, including a spatial frame, a density, a body force, and internal energy. According to this theory, chemical reactions between material constituents are possible, unlike Biot's theory. Although each of the theories has its own application in soft tissue modeling, Cowin and Cardoso (2012) used some of Biot's poroelastic concepts in the mixture theory to model interstitial tissue growth. A good review of poroelastic theory development and fundamentals can be found in Detournay and Cheng (1993).

As Terzaghi and Biot's theories created a scientific revolution in the geotechnical field (de Boer, 1999), they have a great potential to generate a revolution in the field of biomechanics and related medical applications. This can be attributed to the nature of tissues that have a number of common soil characteristics, in that the mechanical behavior of tissues is controlled not only by the solid-phase response, but also by fluids and their movement in and out of the pores. At the instant of loading, the poroelastic material behaves as an elastic solid controlled by shear or elastic modulus of the matrix. This is followed by the fluid flow that continues until equilibrium is reached between the internal fluid pressure and external environment. The pore fluid contributes to the tissue's mechanical behavior by increasing stiffness through pore pressure, and to deformation by its movement within and out of the tissues.

Two scenarios of fluid movement inside the porous media are considered: (1) drained and (2) undrained conditions. In the drained conditions, the pores are assumed to be connected with each other allowing the fluid to move out under loading, and therefore the pore pressure is zero. On the other hand, under the undrained condition, the pores are not connected, and the fluid stays within the pore and contributes to carrying the pressure exerted by external loading. It is worth noting that the fluid contribution to tissue deformation is more pronounced when it flows out of the tissue in drained conditions.

In addition, the response of porous materials can be considered at both macromechanical (continuum) and micromechanical levels. The macromechanical approach describes the overall behavior of a material without describing the contributions of individual constituents by using bulk material properties of K, K_u, and α, representing the bulk modulus of drained elastic solid, the undrained bulk modulus, and the Biot's coefficient, respectively. The Biot's coefficient is the ratio of the gained (or lost) fluid volume of an element to the total volume change of that element when pore pressure is allowed to return to its original state. On the other hand, the micromechanical approach takes the behavior of individual components of a material into consideration

Two strain components are considered to model porous materials behavior: (1) solid and (2) fluid phase-related strains. The well-known solid-related strain (ε_{ij}) is often small and it is positive for extension. The fluid strain component (ζ) represents the fluid content variation (positive for gain of fluid), which is the variation of fluid volume per unit volume of porous material. In the continuum model, Biot's theory assumes a linear relationship between the applied stress (σ_{ij}, p) and strains (ε_{ij}, ζ), in addition to the *elastic* assumption (i.e., full reversibility of deformation) (Detournay and Cheng 1993). To express these parameters in terms of material constants such as K, K_u, and α, the following relationships have been used:

$$\varepsilon = -\frac{1}{K}\left(P - \alpha p\right) \tag{1.13}$$

$$\xi = -\frac{\alpha}{K}\left(P - \frac{p}{B}\right) \tag{1.14}$$

$$B = \frac{K_u - K}{\alpha K_u}$$

where:
P is the total mean isotropic pressure
p is the pore pressure
K and K_u are the drained and undrained bulk moduli, respectively
ε and ζ are the volumetric strain and fluid content variation, respectively

For the volumetric response of elastic isotropic porous materials, drained and undrained conditions are frequently considered. In the undrained condition, the fluid is trapped in the porous media ($\zeta = 0$), the pore pressure is proportional to the total pressure ($p = PB$). The best representation of this case is the instantaneous behavior of the material to a sudden application of loading, where the pore fluid is trapped and has no time to move to other locations except within pores.

However, as the time of loading is extended longer, the fluid has the opportunity to move to other locations within the material and reaches equilibrium within the boundary conditions. For example, if the pressure in the boundary of a material is zero, draining occurs, resulting in total dissipation of the internal pressure ($p = 0$). Therefore, the volumetric strain (ε) is dependent on the mean applied pressure and the drained bulk modulus ($\varepsilon = -P/K$) under drained condition.

The fluid draining out of tissues can be explained using cartilage as an example. It has been observed that the pore fluid can flow through the solid porous matrix of cartilage tissues subjected to pressure, resulting in thickness reduction while acting as a lubricant to reduce friction at joints (Mow et al. 1992). The rate of fluid volume outflow (Q) across a sectional area (A) due to an applied pressure gradient (ΔP) was characterized by the Darcy's law:

$$Q = k\frac{A.\Delta P}{h} \tag{1.15}$$

where k is the hydraulic permeability coefficient and h is the sample thickness. The flow speed (v) is related to the porosity of the tissues ($\varnothing_f$), which is represented by the volumetric ratio of the fluid to the total tissue volume ($\varnothing_f = V_f/V_T$), based on the following equation:

$$v = \frac{Q}{A\varnothing_f} \tag{1.16}$$

The flow resistance is frictional in nature and characterized by the drag coefficient (K), which can be calculated as follows:

$$K = \frac{\left(\varnothing_f\right)^2}{k} \tag{1.17}$$

Mow et al. (1992) reported that the drag coefficient for cartilage was 10^{13} and 10^{14} N s/m^4, which resulted in a high-pressure gradient of 1.5–15 MPa required to move a column of water through a 2 mm thick cartilage specimen at a speed of 10 μm/s. This fluid pressure phenomenon plays a major role in the bearing capacity of cartilage in joints, in addition to its lubrication effect when it flows.

1.5.2 Poroelastic Investigations of Human Tissues

It is known in the origin field of poroelastic theory (geotechnical field) that it is a challenge to find uniformly defined and standard symbols for poroelastic parameters because of the wide range of fields applying the theory (Kümpel 1991). It becomes an even more challenging task to find unified parameters for soft tissues. Therefore, unlike previous sections, numerical parameters of human tissues will be replaced by presenting general poroelastic investigations.

Along with cartilage, brain tissues are among the most investigated tissues using poroelasticity due to the wide range of slow deformation such as the shrinkage of brain after the administration of hyperosmotic drugs to lower intracranial pressure or as a presurgical preparation. For example, Hakim et al. (1976) proposed the original idea of modeling a brain parenchyma as a saturated porous media. This was followed by a wide range of poroelastic models of brain tissues. A number of parameters have been reported; however, few experimental results have supported these poroelasticity models. Franceschini et al. (2006) conducted direct quasi-static uniaxial strain tests using *consolidometer* and showed that the brain tissues obeyed consolidations theory in a manner similar to that of fine soils but with higher volumetric incompressibility.

Although Darcy's law and Terzaghi's linear elasticity have been widely used in a number of poroelasticity applications, the linear elasticity is more appropriate for small deformations. Some soft tissues may experience moderate-to-large deformation in cases such as swelling. Therefore, large deformation poroelasticity using Darcy's law, Terzaghi's stress, and nonlinear elasticity approach becomes more appropriate for soft tissues modeling (MacMinn et al. 2016). Similarly, porohyperelastic modeling was also suggested to capture the shear effect of the blood flow (Suvorov and Selvadurai 2016).

Poroviscoelastic mechanical modeling of soft tissues has been also investigated by a number of researchers. Raghunathan et al. 2010 modeled the liver tissues as poroviscoelastic material to investigate the unconfined compression relaxation under different strain rates (0.001–0.1 s^{-1}). Similarly, Mehrabian (2011) developed a poroviscoelastic model of brain tissues. In some cases, a combination of viscohyperelastic and porous mechanics has been used in modeling soft tissues (Marchesseau et al. 2010).

1.6 Isotropy and Homogeneity of Tissues

The plane of symmetry is the plane through which the material structure has a reflective *mirror* copy of the original. On the other hand, the plane of isotropy can be simply defined as the plane that is created by rotating the material about one axis without noticeable change to its mechanical properties. For example, isotropic material is the material that has the same mechanical properties in all three orthogonal directions (directional based).

Other isotropy-based material types include the following: anisotropic (triclinic), monoclinic (one symmetric plane), orthotropic (three planes of symmetry), transversely isotropic material (one plane of isotropy), and isotropic material (an infinite number of planes of isotropy and symmetry). These classifications are normally used to characterize composite materials such as fiber-reinforced composites; hence, full modeling details of isotropy levels of different materials can be found in composite material books by Daniel and Ishai (1994) and Herakovich (1998). These isotropy models have been used in soft tissue characterizations as they are an example of these composites. On the other hand, a material is called homogeneous if its constituent is the same throughout (location based).

As the solid part of the material included in the above-mentioned models is predominantly represented as linear elastic, the isotropic nature of the solid is discussed here for its simplicity and wide range of applications. In a more generalized case of loading, three stresses are applied on orthogonal directions such as x, y, and z. The stress–strain relationship is established by the Hooke's law, as described in Equation 1.18, for different levels of material isotropy. The number of required constants depends on the type of material investigated. The number of required material constants for each of these materials is listed in Table 1.6.

Although different degrees of anisotropy have been used to model soft tissues (Holzapfel 2001), most of the elastic models concentrate on isotropic modeling due to its simplicity, where only the elastic modulus (E) and Poisson's ratio (ν) are needed.

The generalized Hooke's law developed by Hooke (1678) and Love (1892) is written as follows:

$$\sigma_{ij} = C_{ijkl}\varepsilon_{kl} \tag{1.18}$$

where C_{ijkl} are stiffness coefficients, which are also known as elastic moduli and elastic constants. For *anisotropic* materials, it is a fourth-order tensor, which requires the measurement of 81 (3^4) elastic constants. However, there is no material with that many constants because the symmetry of stress and strain tensors ($\sigma_{ij} = \sigma_{ji}$, $\varepsilon_{ij} = \varepsilon_{ji}$) reduces the parameters to 36 as follows:

$$\sigma_i = C_{ij}\varepsilon_j \quad (i, j = 1, 2 \ldots 6) \tag{1.19}$$

TABLE 1.6 Summary of Materials Symmetry Classification and Required Elastic Parameters

Material Symmetry	Number of Planes of Symmetry	Number of Planes of Isotropy	Number of Nonzero Coefficients	Number of Independent Elastic Constants
Triclinic (anisotropic)	0	0	36	21
Monoclinic	1	0	20	13
Orthotropic	3	0	12	9
Transversely isotropic	$1 + \infty$	1	12	5
Isotropic	∞	∞	12	2

where σ_i and ε_i are the stress and strain components, respectively. The expanded form of Equation 1.19 can be presented as follows:

$$\begin{Bmatrix} \sigma_1 \\ \sigma_2 \\ \sigma_3 \\ \tau_4 \\ \tau_5 \\ \tau_6 \end{Bmatrix} = \begin{bmatrix} C_{11} & C_{12} & C_{13} & C_{14} & C_{15} & C_{16} \\ C_{21} & C_{22} & C_{23} & C_{24} & C_{25} & C_{26} \\ C_{31} & C_{32} & C_{33} & C_{34} & C_{35} & C_{36} \\ C_{41} & C_{42} & C_{43} & C_{44} & C_{45} & C_{46} \\ C_{51} & C_{52} & C_{53} & C_{54} & C_{55} & C_{56} \\ C_{61} & C_{62} & C_{63} & C_{64} & C_{65} & C_{66} \end{bmatrix} \begin{Bmatrix} \varepsilon_1 \\ \varepsilon_2 \\ \varepsilon_3 \\ \gamma_4 \\ \gamma_5 \\ \gamma_6 \end{Bmatrix} \tag{1.20}$$

where:

σ and ε are the normal stress and strain, respectively

τ and γ are the shear stress and shear strain, respectively

Furthermore, the number of required constants can be reduced further due to the symmetry in the stiffness matrix ($C_{ij} = C_{ji}$), which reduces the number of independent elastic constants to 21. The stiffness matrix can be divided into different regions depending on the stress characteristics they represent on a specific plane of the material, as shown in the matrix in Equation 1.21. For example, pure extension (or normal loading) can be captured by three diagonal elements of the matrix (Region I: C_{1}, C_{22}, and C_{33}), whereas pure shear is simulated by the rest of the diagonal matrix elements (C_{44}, C_{55}, and C_{66}). Different coupling between normal–normal, normal–shear, and shear–shear are represented by region II, region III, and region IV, respectively.

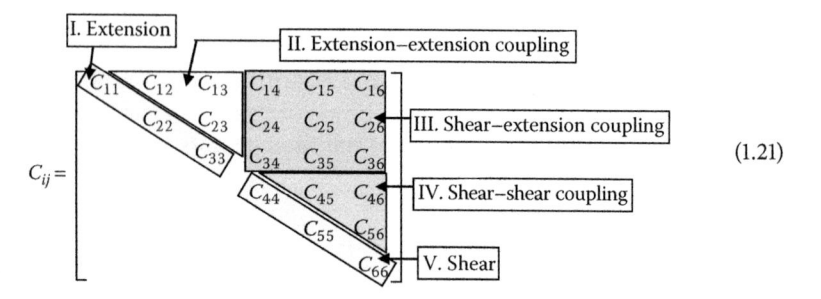

$$\tag{1.21}$$

The most convenient and simple level of material isotropy is the *isotropic* model where only two constants are required, which results in the following stress–strain relationship:

$$\begin{Bmatrix} \sigma_1 \\ \sigma_2 \\ \sigma_3 \\ \tau_4 \\ \tau_5 \\ \tau_6 \end{Bmatrix} = \frac{E}{(1+\upsilon)(1-2\upsilon)} \begin{bmatrix} 1-\upsilon & 0 & 0 & 0 & 0 & 0 \\ 0 & 1-\upsilon & 0 & 0 & 0 & 0 \\ 0 & 0 & 1-\upsilon & 0 & 0 & 0 \\ 0 & 0 & 0 & 1-2\upsilon & 0 & 0 \\ 0 & 0 & 0 & 0 & 1-2\upsilon & 0 \\ 0 & 0 & 0 & 0 & 0 & 1-2\upsilon \end{bmatrix} \begin{Bmatrix} \varepsilon_1 \\ \varepsilon_2 \\ \varepsilon_3 \\ \gamma_4 \\ \gamma_5 \\ \gamma_6 \end{Bmatrix} \tag{1.22}$$

Another familiar representation of the Hooke's law is the inverted form, which can be written as follows:

$$\varepsilon_i = S_{ij}\sigma_j \tag{1.23}$$

where the coefficients S_{ij} are called the compliance coefficients and $S_{ij} = C^{-1}_{ij}$.

For isotropic material, the stress–strain relations are as follows, where only E and v are needed:

$$\begin{Bmatrix} \varepsilon_1 \\ \varepsilon_2 \\ \varepsilon_3 \\ \gamma_4 \\ \gamma_5 \\ \gamma_6 \end{Bmatrix} = \frac{1}{E} \begin{bmatrix} 1 & -v & -v & 0 & 0 & 0 \\ -v & 1 & -v & 0 & 0 & 0 \\ -v & -v & 1 & 0 & 0 & 0 \\ 0 & 0 & 0 & 2(1+v) & 0 & 0 \\ 0 & 0 & 0 & 0 & 2(1+v) & 0 \\ 0 & 0 & 0 & 0 & 0 & 2(1+v) \end{bmatrix} \begin{Bmatrix} \sigma_1 \\ \sigma_2 \\ \sigma_3 \\ \tau_4 \\ \tau_5 \\ \tau_6 \end{Bmatrix} \tag{1.24}$$

1.7 Conclusion

Soft tissues can be mechanically characterized as linear elastic, hyperelastic, viscoelastic, and poroelastic materials. The choice of characterization approach is normally application dependent, where the required outputs in addition to the type of applied load play a major role in the model-selection processes. Linear elastic modeling has been widely used for their simplicity and efficiency in modeling regular physiological activities for applications such as image-guided radiotherapy. On the other hand, hyperelastic modeling can be used to capture a wide range of material behavior in physiological and traumatic loading cases. However, more computing time and details are needed. When time is required as a modeling element of tissue behavior, viscoelastic properties are needed. Similarly, as more detailed outcomes are needed for some applications, tissues can be modeled as poroelastic materials.

References

Al-Mayah, A., and K. Brock. 2013. Toward realistic biomechanical based modeling for image guided radiation. *Imaging in Medical Diagnosis and Therapy: Image Processing in Radiation Therapy*, CRC Press, Taylor & Francis Group, Boca Raton, FL, 85–94.

Asbach, P., D. Klatt, U. Hamhaber, J. Braun, R. Somasundaram, B. Hamm, and I. Sack. 2008. Assessment of liver viscoelasticity using multifrequency MR elastography. *Magn Reson Med* 60: 373–379.

Biot, M. A. 1941. General theory of three-dimensional consolidation. *J Appl Phys* 12(2): 155–164.

Bowen, R. M. 1980. Incompressible porous media models by use of the theory of mixtures. *Int J Eng Sci* 18(9): 1129–1148.

Boyce, M. C. and E. M. Arruda. 2000. Constitutive models of rubber elasticity: A review. *Rubber Chem Technol* 73: 504–523.

Brunon, A., K. Bruyere-Garnier, and M. Coret. 2011. Characterization of the nonlinear behavior and the failure of human liver capsule through inflation tests. *J Mech Behav Biomed Mater* 4(8): 1572–1581.

Budday, S., G. Sommer, C. Birkl, C. Langkammer, J. Haybaeck, J. Kohnert, M. Bauer et al. 2017. Mechanical characterization of human brain tissue. *Acta Biomater* 48: 319–340.

Carter, F. J., T. G. Frank, P. J. Davies, D. McLean, and A. Cuschieri. 2001. Measurement and modelling of the compliance of human and porcine organs. *Med Image Anal* 5: 231–236.

Chew P. H., F. C. Yin, and S. L. Zeger. 1986. Biaxial stress–strain properties of canine pericardium. *J Mol Cell Cardiol* 18(6): 567–578.

Corpechot, C., A. El Naggar, A. Poujol-Robert, M. Ziol, D. Wendum, O. Chazouillères, V. de Lédinghen et al. 2006. Assessment of biliary fibrosis by transient elastography in patients with PBC and PSC. *Hepatology* 43(5): 1118–1124.

Cowin, S. C., and L. Cardoso. 2012. Mixture theory-based poroelasticity as a model of interstitial tissue growth. *Mech Mater* 44: 47–57.

Criscione J. C., M. S. Sacks, and W. C. Hunter. 2003. Experimentally tractable, pseudoelastic constitutive law for biomembranes: I. Theory. *J Biomech Eng* 125(1): 94–99.

Daniel, I. M., and O. Ishai. 1994. *Engineering Mechanics of Composite Materials.* New York: Oxford university press.

de Boer, R. 1999. *Theory of Porous Media.* Springer, Berlin, Germany.

Detournay, E., and H. D. A. Cheng. 1993. Fundamentals of poroelasticity. In: Hudson, J. A. (Ed.), *Comprehensive Rock Engineering: Principles, Practice and Projects.* Pergamon, Oxford, UK, Vol 2: 113—171.

Dolhnikoff, M., J. Morin, and M. S. Ludwig. 1998. Human lung parenchyma responds to contractile stimulation. *Am J Respir Crit Care Med* 158(5): 1607–1612.

Franceschini, G., D. Bigoni, P. Regitnig, and G. A. Holzapfel. 2006. Brain tissue deforms similarly to filled elastomers and follows consolidation theory. *J Mech Phys Solids* 54(12): 2592–2620.

Fung, Y. C. 1975. Stress, deformation, and atelectasis of the lung. *Circ Res* 37(4): 481–496.

Fung, Y. C. 1993. *Biomechanics: Mechanical Properties of Living Tissues.* Springer, New York.

Gao, J., W. Huang, and R. Yen. 2006. Mechanical properties of human lung parenchyma. *Biomed Sci Instrum* 42: 172.

Gardiner, J. C., and J. A. Weiss 2001. Simple shear testing of parallel fibered planar soft tissues. *J Biomech Eng* 123(2): 170–175.

Green, M. A., L. E. Bilston, and R. Sinkus. 2008. In vivo brain viscoelastic properties measured by magnetic resonance elastography. *NMR Biomed* 21(7): 755–764.

Hakim, S., J. G. Venegas, and J. D. Burton. 1976. The physics of the cranial cavity, hydrocephalus and normal pressure hydrocephalus: Mechanical interpretation and mathematical model. *Surg Neurol* 5(3): 187–210.

Herakovich, C. T. 1998. *Mechanics of Fibrous Composites.* John Wiley & Sons, New York.

Holzapfel, G. A. 2001. Biomechanics of soft tissue. *The Handbook of Materials Behavior Models* 3: 1049–1063.

Humphrey J. D., R. K. Strumpf, and F. C. Yin. 1990a. Determination of a constitutive relation for passive myocardium: I. A new functional form. *J Biomech Eng* 112(3): 333–339.

Humphrey J. D., R. K. Strumpf, and F. C. Yin. 1990b. Determination of a constitutive relation for passive myocardium: II. Parameter estimation. *J Biomech Eng* 112(3): 340–346.

Humphrey, J. D. 2003. Continuum biomechanics of soft biological tissues. *Proceedings of the Royal Society of London. Series A*, Mathematical and Physical Sciences 459: 3–4.

Jarrahi, A., A. Karimi, M. Navidbakhsh, and H. Ahmadi. 2016. Experimental/numerical study to assess mechanical properties of healthy and Marfan syndrome ascending thoracic aorta under axial and circumferential loading. *Mater Tech* 31(5): 247–254.

Karimi, A., M. Navidbakhsh, A. Shojaei, and S. Faghihi. 2013. Measurement of the uniaxial mechanical properties of healthy and atherosclerotic human coronary arteries. *Mater Sci Eng C Mater Biol Appl* 33(5): 2550–2554.

Klatt, D., U. Hamhaber, P. Asbach, J. Braun, and I. Sack. 2007. Noninvasive assessment of the rheological behavior of human organs using multifrequency MR elastography: A study of brain and liver viscoelasticity. *Phys Med Biol* 52(24): 7281.

Krouskop, T. A., T. M. Wheeler, F. Kallel, B. S. Garra, and T. Hall. 1998. Elastic moduli of breast and prostate tissues under compression. *Ultrason Imaging* 20(4): 260–274.

Kümpel, H. J. 1991. Poroelasticity: Parameters reviewed. *Geophys J Int* 105(3): 783–799.

Lai-Fook, S. J. and R. E. Hyatt. 2000. Effect of age on elastic moduli of human lungs. *J Appl Physiol* 89: 163–168.

Lyshchik, A., T. Higashi, R. Asato, S. Tanaka, J. Ito, M. Hiraoka, A. B. Brill, T. Saga, and K. Togashi. 2005. Elastic moduli of thyroid tissues under compression. *Ultrason Imaging* 27(2): 101–110.

MacMinn, C. W., E. R. Dufresne, and J. S. Wettlaufer. 2016. Large deformations of a soft porous material. *Phys Rev Applied* 5(4): 044020.

Marcellin, P., M. Zoil, P. Bedossa, C. Douvin, R. Poupon, V. De Lédinghen, and M. Beaugrand. 2009. Non-invasive assessment of liver fibrosis by stiffness measurement in patients with chronic hepatitis B. *Liver Int* 29: 242–247.

Marchesseau, S., T. Heimann, S. Chatelin, R. Willinger, and H. Delingette. 2010. Fast porous visco-hyperelastic soft tissue model for surgery simulation: Application to liver surgery. *Prog Biophys Mol Biol* 103(2): 185–196.

Marckmann, G., and E. Verron. 2006. Comparison of hyperelastic models for rubberlike materials. *Rubber Chem Technol* 79(5): 835–858.

Mariappan, Y. K., K. J. Glaser, R. D. Hubmayr, A. Manduca, R. L., Ehmanand, and K. P. McGee. 2011. MR elastography of human lung parenchyma: technical development, theoretical modeling and in vivo validation. *J Magn Reson Imaging* 33(6): 1351–1361.

Martins, P. A. L. S., R. M. Natal Jorge, and A. J. M. Ferreira. 2006. A comparative study of several material models for prediction of hyperelastic properties: Application to silicone-rubber and soft tissues. *Strain* 42: 135–147.

Mehrabian, A., and Y. Abousleiman. 2011. General solutions to poroviscoelastic model of hydrocephalic human brain tissue. *J Theor Biol* 291: 105–118.

Mow, V. C., A. Ratcliffe, and A. R. Poole. 1992. Cartilage and diarthrodial joints as paradigms for hierarchical materials and structures. *Biomaterials* 13(2): 67–97.

Muller, M., J. L. Gennisson, T. Deffieux, M. Tanter, and M. Fink. 2009. Quantitative viscoelasticity mapping of human liver using supersonic shear imaging: Preliminary in vivo feasibility study. *Ultrasound Med Biol* 35: 219–229.

Nava, A., E. Mazza, M. Furrer, P. Villiger, and W. H. Reinhart. 2008. In vivo mechanical characterization of human liver. *Med Image Anal* 12(2): 203–216.

Ogden, R. W. 1984. *Non-Linear Elastic Deformations*. Dover Publications, New York.

O'Hagan, J. J., and A. Samani. 2009. Measurement of the hyperelastic properties of 44 pathological ex vivo breast tissue samples. *Phys Med Biol* 54(8): 2557.

Raghunathan, S., D. Evans, and J. L. Sparks. 2010. Poroviscoelastic modeling of liver biomechanical response in unconfined compression. *Ann Biomed Eng* 38(5): 1789–1800.

Rice, J. R. and M. P. Cleary. 1976. Some basic stress-diffusion solutions for fluid saturated elastic porous media with compressible constituents. *Rev Geophys Space Phys* 14: 227–241.

Samani, A., and D. Plewes. 2004. A method to measure the hyperelastic parameters of ex vivo breast tissue samples. *Phys Med Biol* 49(18): 4395.

Schiavone, P., F. Chassat, T. Boudou, E. Promayon, F. Valdivia, and Y. Payan. 2009. In vivo measurement of human brain elasticity using a light aspiration device. *Med Image Anal* 13(4): 673–678.

Sinkus, R., M. Tanter, T. Xydeas, S. Catheline, J. Bercoff, and M. Fink. 2005. Viscoelastic shear properties of in vivo breast lesions measured by MR elastography. *Magn Reson Imaging* 23(2): 159–165.

Sun, W., M. S. Sacks, and M. J. Scott. 2005. Effects of boundary conditions on the estimation of the planar biaxial mechanical properties of soft tissues. *J Biomech Eng* 127(4): 709–715.

Suvorov, A. P., and A. P. S. Selvadurai. 2016. On poro-hyperelastic shear. *J Mech Phys Solids* 96: 445–459.

Takeda, T., G. Kassab, J. Liu, J. L. Puckett, R. R. Mittal, and R. K. Mittal. 2002. A novel ultrasound technique to study the biomechanics of the human esophagus in vivo. *Am J Physiol Gastrointest Liver Physiol* 282(5): G785–G793.

Terzaghi, K. 1925. Principles of soil mechanics, IV—Settlement and consolidation of clay. *Eng News-Rec* 95(3): 874–878.

Tong, P., and Y. C. Fung. 1976. The stress–strain relationship for the skin. *J Biomech* 9(10): 649–657.

Truesdell, C., and R. A. Toupin. 1960. The classical field theories of mechanics. In S. Flügge (Ed.), *Handbook of Physics*. Springer, New York.

Uffmann K., S. Maderwald, A. de Greiff, and M. Ladd. 2004. Determination of gray and white matter elasticity with MR elastography. *Proceedings of the 12th Annual Meeting of ISMRM (Kyoto)*. p. 1768.

Van Houten, E. E., M. M. Doyley, F. E. Kennedy, J. B. Weaver, and K. D. Paulsen. 2003. Initial in vivo experience with steady-state subzone-based MR elastography of the human breast. *J Magn Reson Imaging* 17(1): 72–85.

Verruijt, A., 1969. The completeness of Biot's solution of the coupled thermoelastic problem. *Q Appl Math* 26: 485–490.

Weiss J. A., B. N. Maker, and S. Govindjee. 1996. Finite element implementation of incompressible, transversely isotropic hyperelasticity. *Comput Methods Biomech Biomed Eng* 135: 107–128.

Wierzbicka, M., Kałużny, J., Szczepanek-Parulska. E., A. Stangierski, E. Gurgul, T. Kopeć, and M. Ruchała. 2013. Is sonoelastography a helpful method for evaluation of parotid tumors? *Eur Arch Otorhinolaryngol* 270(7): 2101–2107.

Yeh, W. C., P. C. Li, Y. M. Jeng, H. C. Hsu, P. L. Kuo, M. L. Li, P. M. Yang, and P. H. Lee. 2002. Elastic modulus measurements of human liver and correlation with pathology. *Ultrasound Med Biol* 28: 467–474.

Zeng, Y. J., D. Yager, and Y. C. Fung. 1987. Measurement of the mechanical properties of the human lung tissue. *J Biomech Eng* 109: 169–174.

Zhang, J., M. A. Green, R. Sinkus, and L. E. Bilston. 2011. Viscoelastic properties of human cerebellum using magnetic resonance elastography. *J Biomech* 44(10): 1909–1913.

Zhang, M., S. Fu, Y. Zhang, J. Tang, and Y. Zhou. 2014. Elastic modulus of the prostate: A new non-invasive feature to diagnose bladder outlet obstruction in patients with benign prostatic hyperplasia. *Ultrasound Med Biol* 40(7): 1408–1413.

Zhang, M., P. Nigwekar, B. Castaneda, K. Hoyt et al. 2008. Quantitative characterization of viscoelastic properties of human prostate correlated with histology. *Ultrasound Med Biol* 34(7): 1033–1042.

2

Mechanical Investigations of Biological Tissues Using Tensile Loading and Indentation

Wanis Nafo and
Adil Al-Mayah

2.1 Introduction

In the field of biomechanics, it is important to understand the mechanical properties of biological tissues in order to model their behavior under normal and abnormal pathological and traumatic conditions. Biomechanical modeling of soft tissues can play a significant role in improving a wide range of medical procedures, including diagnostics, treatment, planning, and interventions (Holzapfel 2004). In addition, studying the mechanics of biological tissues is essential to predict their response in traumatic incidents of extreme loading conditions such as vehicle and sports accidents.

Tensile and indentation-testing techniques are fundamental approaches to characterize the behavior of a wide range of materials. These tests are based on applying loads to material samples while monitoring corresponding deformation, sometimes to failure, in an attempt to understand the mechanical behavior of a material. They have been widely used in the field of biomedical engineering in which biological tissues are deformed with certain strain rates to characterize their mechanical response.

Biological soft tissues are inhomogeneous by nature and tend to behave anisotropically under loading, which poses a challenge to understanding their mechanics. As a result, most of the available literature concentrates on the overall mechanical performance by assuming that tissues behave isotropically under loading using techniques such as tensile and indentation loading. This chapter presents an overview of these studies and introduces the fundamentals of indentation and tensile testing. Examples of how different biological tissue morphologies impact tissue mechanics are also presented.

2.2 Tensile Test

Tensile testing enables direct measurement of the mechanical properties of *ex vivo* soft tissues. The load versus extension data obtained from these tests are used to calculate stress versus deformation relationships. There are two common types of tensile tests: (1) uniaxial and (2) biaxial. Uniaxial tests are commonly used when material properties are considered the same in all directions, whereas biaxial procedures are important when material properties are direction-dependent. As a result, uniaxial tests stretch material samples in a single direction, whereas biaxial tests stretch them in two perpendicular directions. The common sample shapes used in uniaxial and biaxial testing are shown in Figure 2.1a and b, respectively.

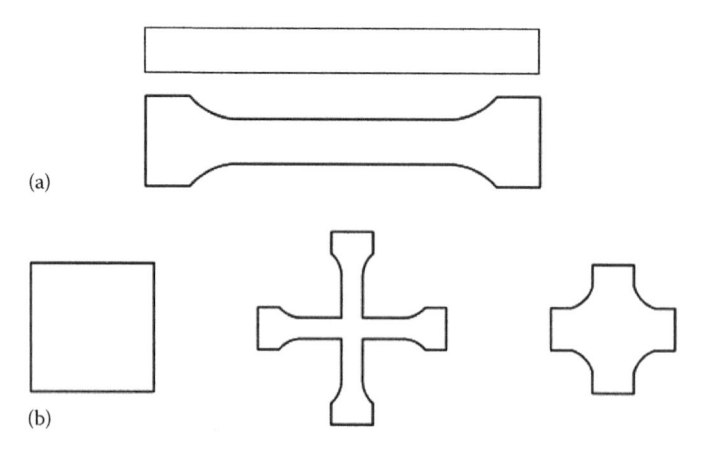

FIGURE 2.1 Common sample shapes used in (a) uniaxial tensile tests and (b) biaxial tensile tests.

2.2.1 Uniaxial Tensile Test

Uniaxial tensile tests are among the most widely used methods for characterizing the mechanical response of soft tissues to mechanical loading, owing to relatively simple procedures and direct test data analysis (Jacobs et al. 2013). The common applications are testing fibrous connective tissues such as tendons and ligaments that connect muscles to bones and bones-to-bones, respectively. These tissues are composed of aligned collagen fibers that make very strong and small fibrils, and provide load-bearing capacity in one direction (for the most part), normally parallel to the longitudinal axis of connective tissues (Lynch 2003). In general, the uniaxial tensile test represents a good option to capture the deformation mechanics of these tissues.

As mentioned previously, uniaxial tests are relatively simple to perform. A sample with known length L and cross-sectional area A is gripped at both the ends, mounted on a loading machine, and stretched at a predetermined rate. Special attention must be paid to the gripping force applied to the sample where high-gripping stress may result in unwanted high stresses that rupture the sample prematurely. On the other hand, low gripping causes the sample to slip out of the grip. Introducing rough surfaces, such as sand paper, increases the friction between the sample and grip device can reduce, and even prevent, slippage (Kemper et al. 2010).

A controlled stretching load F is applied at a specific rate along the sample throughout the test. As the load increases gradually, the sample deforms accordingly. The load–deformation relationship can be normalized to derive the stress–strain (σ–ε) relationship or stress–stretch (σ–λ) relationship, which is then used to extract parameters used in a variety of constitutive equations that describe the mechanical behavior of materials. Figure 2.2 illustrates the deformation process.

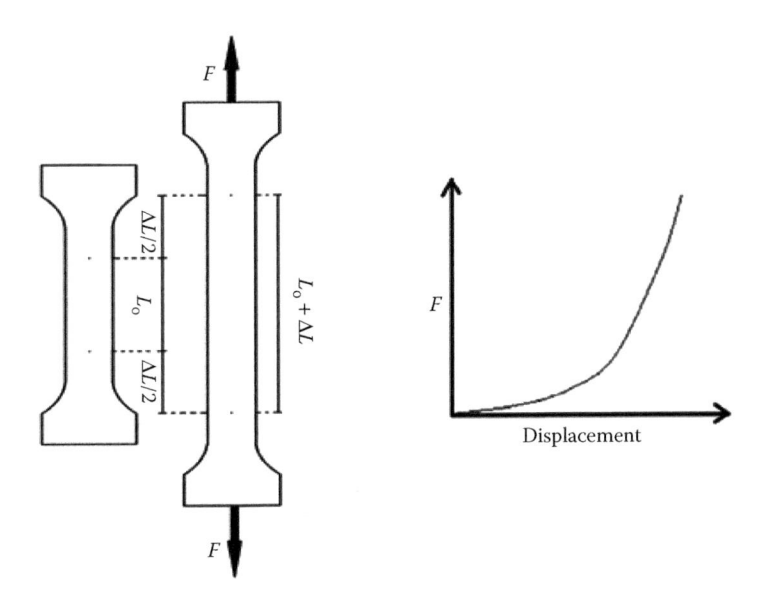

FIGURE 2.2 Uniaxial stretching.

Stress (σ) is defined as the load (F) divided by the cross-sectional area (A) of the sample, as follows:

$$\sigma = \frac{F}{A} \tag{2.1}$$

Strain (ε) is a ratio of elongation (ΔL) to the original length (L_0), where

$$\varepsilon = \frac{\Delta L}{L_0} \tag{2.2}$$

However, stretch (λ) can be defined as the ratio of the deformed length ($L_0 + \Delta L$) to the original length (L_0), where

$$\lambda = \frac{L_0 + \Delta L}{L_0} \tag{2.3}$$

For a linear elastic isotropic material, stresses and strains can be correlated using the following relationship:

$$\sigma = E.\varepsilon \tag{2.4}$$

where E is the elastic modulus and represents the material stiffness.

2.2.2 Biaxial Tensile Test

In general, due to the high mechanical anisotropy of some soft tissues, the uniaxial tensile test is not sufficient for reliable characterization of parameters associated with the constitutive models used to predict the tissue's behavior. Even when the uniaxial test is applied in many directions to obtain multidirectional data, results cannot be implemented in generalized three-dimensional (3D) constitutive models (Sacks 2000). Therefore, the biaxial tensile test was introduced to provide a more accurate mechanical characterization of soft tissues. The applied deformation in each direction during biaxial tensile tests can be controlled independently or coordinated to a prescribed stretching regime. Figure 2.3 shows a biaxial test setup in which a sample is subjected to perpendicular stretching. This feature provides the ability to observe the mechanical response of the biological samples in various stress states.

For biaxial tests, the simplest forms of stresses and strains are defined as follows:

$$\varepsilon = \begin{bmatrix} \varepsilon_1 & \varepsilon_{12} & 0 \\ \varepsilon_{21} & \varepsilon_2 & 0 \\ 0 & 0 & \varepsilon_3 \end{bmatrix}, \quad \sigma = \begin{bmatrix} \sigma_1 & \sigma_{12} & 0 \\ \sigma_{21} & \sigma_2 & 0 \\ 0 & 0 & 0 \end{bmatrix} \tag{2.5}$$

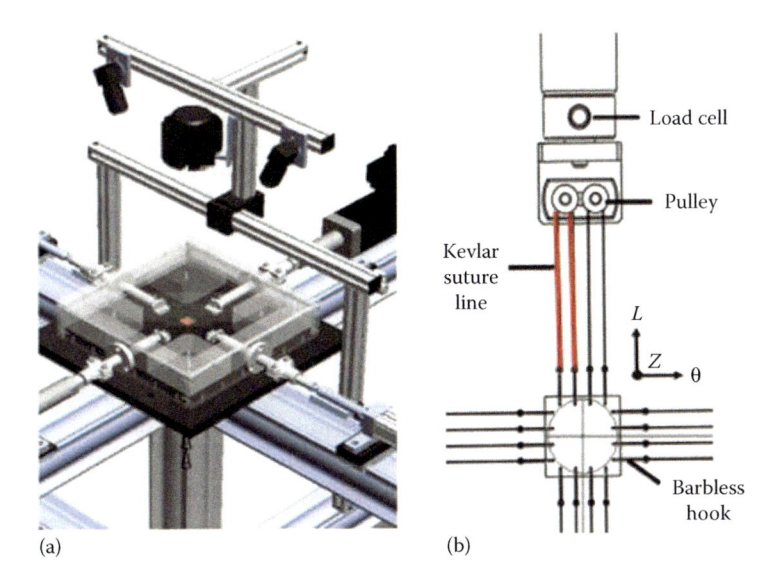

(a)   (b)

FIGURE 2.3 (a) Biaxial tensile test experimental setup and (b) top view schematic of the test setup. (Reprinted from Deplano, V. et al., *J. Biomech.*, 49, 2031–2037, 2016. With permission.)

where:

σ_1 and σ_2 represent stresses in the directions of the applied loads

ε_1 and ε_2 are the corresponding strains

σ_{12} and σ_{21} are shear stresses

ε_{12} and ε_{21} are the corresponding shear strains

ε_3 is commonly known as the thickness strain

The nonzero value of this strain is due to Poisson's ratio (υ) effect. The relationships between stresses and strains are derived as follows:

$$\varepsilon_1 = \frac{1}{E}\left(\sigma_1 - \upsilon\sigma_2\right)$$

$$\varepsilon_2 = \frac{1}{E}\left(\sigma_2 - \upsilon\sigma_1\right) \qquad (2.6)$$

$$\varepsilon_3 = -\frac{1}{E}\left(\upsilon\sigma_1 + \upsilon\sigma_2\right)$$

$$\varepsilon_{12} = \frac{\sigma_{12}}{G}, \; \varepsilon_{21} = \frac{\sigma_{21}}{G}$$

where G is the shear or rigidity modulus.

The biaxial tensile test can be categorized based on three stress states. In the first state, called equibiaxial tension, loads are applied to maintain equal deformations in both directions throughout the test. The second state is achieved by applying proportional tension where proportional deformations for both directions are maintained throughout the test. In the third state, a constant transverse deformation is applied by increasing the tensile load in one direction, whereas a zero contraction is maintained in the transverse direction, commonly known as the planar tension test (Zemanek et al. 2009).

The variety of stress states enables more accurate measurement of parameters used in constitutive models that predict tissue behavior. This has been observed in annulus fibrosis tissue testing, where uniaxial and biaxial methods produced significantly different results. On the basis of biaxial tensile tests, Gregory and Callaghan (2011) reported that the maximum stress and stress–stretch ratio were 97% and 118%, respectively; much higher than the parameters obtained from the uniaxial loading of annulus fibrosis deformed at 1.23 stretch ratio (Gregory and Callaghan 2011). In general, the differences in the mechanical behavior of any material are based on the boundary conditions imposed by testing methods. In uniaxial tensile tests, the lateral deformation is not restrained, which is reflected in relatively high deformations when the applied stresses are relatively low. On the other hand, materials can experience considerably higher stresses at lower strains because the lateral deformation is confined either by geometry (high width to length ratio) in planar tension tests or by the two-dimensional (2D) load application in biaxial tensile test. An example of these differences is illustrated in Figure 2.4.

It is worth mentioning that when a material sample is tested using conventional mechanical testing techniques such as stretching or compression, the analysis of the stresses experienced by the sample is traditionally based on uniformly distributed stresses on a volume equivalent to the product of the gage length (l_0) and cross-sectional area (A_0) represented by the shaded area shown in Figure 2.5. However, nonuniform stress distribution complications are common in these two techniques. In tension, the gripping system

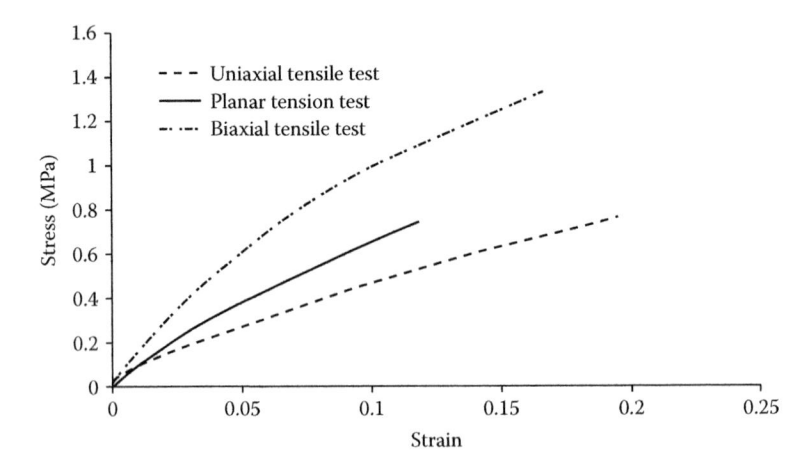

FIGURE 2.4 Stress–strain relationship of different types of tensile test. (Data from Duncan, B.C. et al., Verification of Hyperelastic Test Methods PAJ1 Report 17, 1999.)

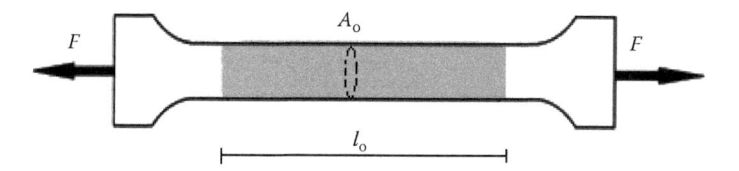

FIGURE 2.5 Illustration of the tested volume (V) relative to the sample's volume ($V_0 = A_0.l_0$) in uniaxial tensile test.

that used to clamp the sample plays a significant role in stress distribution inside the tested sample (Mucsi 2013). In the compression test, the friction that usually occurs between the surface of the sample and the testing rig can cause nonuniformity of stress distribution through the gauge length of the material (Blake 1985). Therefore, other techniques with a more localized loading distribution, such as indentation, may be more suitable for testing soft tissue properties.

2.3 Indentation

Indentation tests are similar to palpation tests, the oldest diagnostic technique, which qualitatively measures tissue stiffness based on the deformation response to an applied hand pressure. However, indentation is a quantitative technique where the applied force and the corresponding deformation (displacement) are measured using force and displacement sensors. These measured data are then used to extract mechanical parameters of soft tissues, including elastic modulus, shear modulus, and hyperelastic parameters associated with energy functions.

One of the main attractions of the indentation technique is its simplicity. In addition, it represents a very good option for measuring the mechanical properties of local heterogeneities in biological materials (Oyen 2011), where constituents have multiscale organizations. For example, tendons have multiscale organizations that range from the subnanostructure level (collagen molecules) to macrostructure levels (fascicle) (Fratzl and Weinkamer 2007). Furthermore, as the indentation test is based on localized contact mechanics at the indenter–material interface, it is not subject to complications arising from gripping effects that is common in tensile testing. Therefore, the tested volume of the sample is small by comparison with the gage volume used in tensile test.

The mechanical response of the material depends fundamentally on the geometry of the indenter (Oyen 2011). Different geometries of indentation tips have been used, including spherical, conical (or pyramidal), and flat-punch (cylindrical or flat-end conical) tips (Figure 2.6). The spherical indenter is controlled by two main parameters: (1) the radius of the sphere (R) and the (2) contact radius (a). In conical or pyramidal tips (sharp tips), the tips form pointed ends that are controlled by the effective angle (α) of the tip. The geometry of the indenter in the flat punch at the contact region is characterized by the diameter (D) of the indenter tip.

The selection of indenter type depends mainly on the type of data required. For instance, the pointed end of a sharp indenter concentrates the applied loads on a very small area, initiating an almost immediate plastic response (Atkins and Tabor 1965, Oyen 2011).

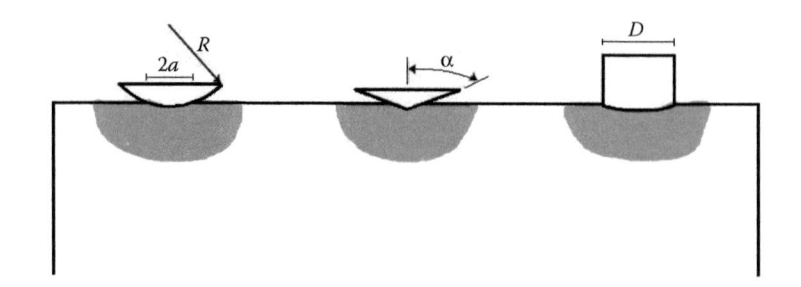

FIGURE 2.6   Schematic illustration of common indentation tips.

On the other hand, flat and cylindrical tips are relatively large in terms of the contact area, and are therefore commonly used to test soft materials (Ebenstein and Pruitt 2004, Liu et al. 2009, Tong and Ebenstein 2015).

Despite the simplicity of defining the contact area in the case of a flat tip, the sharp edges of the tip generate localized stress concentrations (Liu et al. 2009). Although this problem is eliminated when using a spherical tip, it is challenging to calculate the contact area as the load increases (Ebenstein and Pruitt 2004, Tong and Ebenstein 2015). The following section highlights the analytical procedure differences between these types of indenters.

2.3.1 Spherical Indentation

The indentation technique is based on the contact mechanics pioneered by Heinrich Hertz (1881). In his work, the effective radius (R_e) of two contacting elastic spherical bodies with radii R_1 and R_2 can be calculated as follows:

$$\frac{1}{R_e} = \frac{1}{R_1} + \frac{1}{R_2} \tag{2.7}$$

Similarly, the combined modulus (E_r) of two contacting bodies with elastic moduli of E and E' and Poisson's ratios of υ, and υ' is defined as follows (Timoshenko and Goodier 1951):

$$\frac{1}{E_r} = \frac{1-\upsilon^2}{E} + \frac{1-\upsilon'^2}{E'} \tag{2.8}$$

In indentation, Hertz contact theory is adopted in scenarios where a spherical indenter penetrates a flat material surface, as illustrated in Figure 2.7. The theory indicates that the radius of contact circle a is dependent on the indenter geometry (R), the force applied by the indenter (P), and the material's elastic properties as follows (Fischer-Cripps 2007):

$$a^3 = \frac{3}{4}\frac{PR}{E_r} \tag{2.9}$$

where E_r is the reduced elastic modulus.

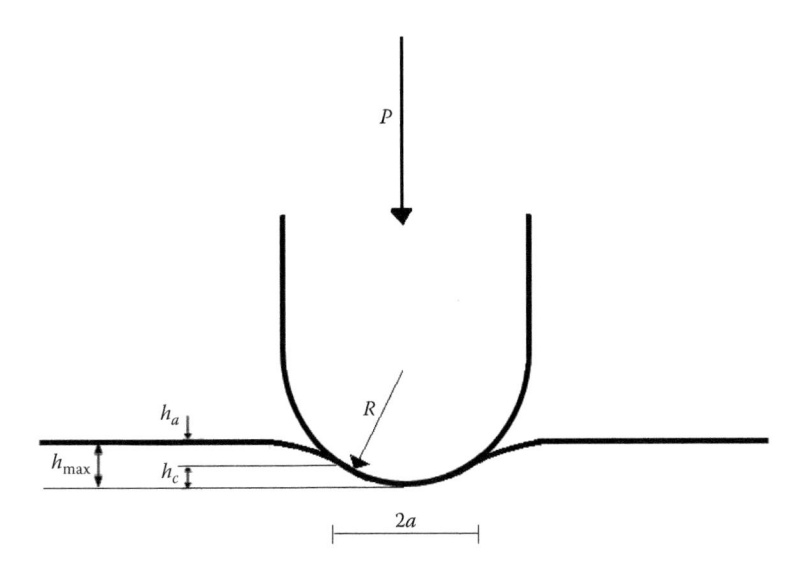

FIGURE 2.7 Illustration of the contact between a rigid indenter and an elastic material: h_{max} is the total depth of penetration, h_a is the depth of the contact region from the surface of the material, and h_c is the contact depth.

When the load is applied, the indenter penetrates the surface by a traveling distance of (h). Figure 2.7 provides an illustration of this contact, including the total depth of penetration (h_{max}), known as the *load-point displacement*, defined by Fischer-Cripps (2011) as follows:

$$h^3 = \left(\frac{3}{4 E_r}\right)^2 \frac{P^2}{R}$$ (2.10)

Indentation test analysis depends on the rigidity of the indenter: rigid or nonrigid. For *rigid indentation*, the indenter penetrates a specimen of modulus E. The indenter is significantly stiffer than the material ($E_{indenter} \gg E_{sample}$); therefore, the reduced modulus E_r becomes modulus of the material and redefined as follows (Oyen 2011):

$$E_r = \frac{E}{\left(1 - \upsilon^2\right)}$$ (2.11)

where E and υ are the elastic modulus and Poisson's ratio, respectively, of the investigated material.

For the nonrigid indentation, the analysis must make one of the two assumptions, based on Fischer-Cripps (2001). Specifically, Equation 2.8 allows the analysis to be treated as (1) a contact between nonrigid indenter of modulus E' and a specimen of modulus E or (2) as a contact between a rigid indenter of radius R_i and a specimen with modulus E_r.

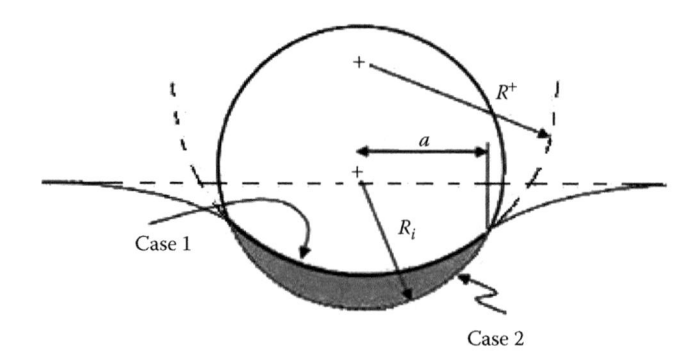

FIGURE 2.8 Illustration of indentation analysis assumptions. Case 1 represents a nonrigid indenter penetrating a material surface of modulus E. Case 2 represents a presumed penetration of a rigid indenter into material of modulus E_r. The indenter does not physically penetrate the material in Case 2 (gray area); therefore, in the analysis it can be assumed that a rigid indenter of radius R^+ penetrates a material of modulus E. (Reprinted from Fischer-Cripps, A.C. *J. Mater. Res.*, 16, 3050–3052, 2001. With permission.)

In the latter case, h becomes the total displacement resulting from indentation h_{max} (Fischer-Cripps 2011), where

$$P = \frac{4}{3} E_r R^{\frac{1}{2}} h^{\frac{3}{2}} \tag{2.12}$$

Figure 2.8 shows the two cases of nonrigid spherical indentation.

However, one must realize that in this case, the physical deformation experienced by the indented material surface is analyzed as an indentation by an indenter of larger radius (R^+), that is, (R) in Equation 2.12 becomes R^+, which is defined as follows (Fischer-Cripps 2011):

$$R^+ = \frac{4a^3 E}{3\left(1-\upsilon^2\right)P} \tag{2.13}$$

where E and υ are the material elastic modulus and Poisson's ratio, respectively.

Some concerns were raised by Chaudhri (2001) regarding the assumption of reduced elastic modulus in the case of nonrigid indenter (Chaudhri 2001). He argued that using the reduced modulus does not have any *theoretical justification*, which he then illustrated using two mathematical examples that led to unreal results. Therefore, it was suggested that the indenter should have a very high elastic modulus compared to the indented material as long as there is no analytical solution that precisely interprets the force–displacement data obtained from the test. These concerns were addressed analytically in Fischer-Cripps (2003), and Cao et al. (2007).

2.3.2 Flat-Punch and Sharp Elastic Indentations

Boussinesq (1885) pioneered early studies on stress distributions in contact problems where an elastic material surface (half space) is subjected to pressure by a rigid punch obtained by rotating a plane around an axis. This axis is normal to the undeformed boundary of the half space (Sneddon 1965), where, for example, an elastic material with flat surface is punched by either a cone or cylinder. More details of these indenters are presented in Sneddon (1946, 1948).

In flat-punch indentation, shown in Figure 2.9, the load–displacement relationship can be defined as follows (Sneddon 1965):

$$P = \frac{2aEh}{1 - \upsilon^2} \tag{2.14}$$

where a, E, υ, and h are the contact radius, elastic modulus, Poisson's ratio of the material, and indentation depth, respectively.

In contrast to spherical indenters, the depth of penetration caused by the cylindrical indenter is the same in cases of rigid and nonrigid indenters because the contact radius (a) is constant. Therefore, the applied pressure is the same in both cases. If the material is elastic, using a cylindrical indenter will result in a linear load–displacement relationship; on the other hand, a spherical indenter will result in a relatively nonlinear load–displacement relationship (Fischer-Cripps 2007).

In sharp indentation, a sharp cone-like indenter is commonly used, as shown in Figure 2.10, and the load–displacement relationship can be defined as follows (Sneddon 1965):

$$P = \frac{2E\tan\alpha}{\pi\left(1 - \upsilon^2\right)} h^2 \tag{2.15}$$

where α, E, v, and h are the half angle of the cone-like indenter, Young's modulus of the material, Poisson's ratio, and the indentation depth, respectively.

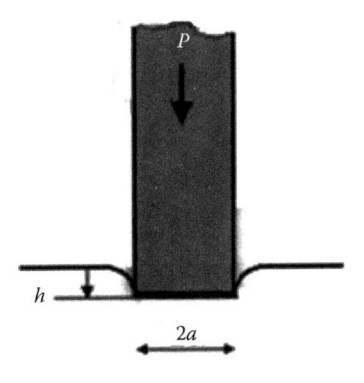

FIGURE 2.9 Schematic illustration of flat-punch indentation.

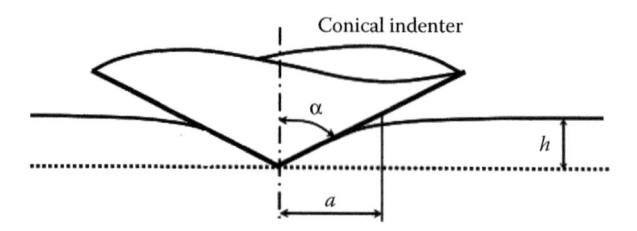

FIGURE 2.10 Geometrical illustration of sharp indentation. (Reprinted from Gouldstone, A. et al., *Acta Materialia*, 55, 4015–4039, 2007. With permission.)

The sharp contact shares a geometrical property with spherical contacts, in that the ratio of the radius of the contact area to the penetration depth (a/h) remains the same under loading. However, the nature of this ratio differs between the two contact types. In spherical indentation, when loads are applied, the penetration depth (h) increases at a slower rate than the radius of the contact circle (Fischer-Cripps 2011). On the other hand, the angle (α) of the tip controls the magnitude of the ratio between (a) and (h).

2.4 Morphology and Mechanical Properties of Biological Tissues

2.4.1 General

Given the wide variation in structural organization and the biological nature of soft biological tissues, their response to mechanical loading is expected to be different as well. In addition, the testing method, loading direction, and loading rate can influence the measurement of the mechanical properties of the same tissue. Therefore, mechanical investigations of several different tissues are presented in the following sections, including musculoskeletal tissues and internal organs such as the brain, liver, and spleen.

2.4.2 Musculoskeletal Tissues

The fibrous structure of articular tissue collagen fibers, such as in tendons, ligaments, and skeletal muscles, is shown to be crimped in their unloaded phase (Dale et al. 1972). However, when these structures are subjected to tensile loading, their response can be divided into three main stages. First, at the early stage of loading ($\varepsilon < 3\%$), fibers show very small resistance as they disentangle their waviness and, as the load increases, more fibers are recruited to resist the load. This mechanism results in a nonlinear response known as the *toe region*. Second, when all the fibers are fully extended, the fibers tend to deform linearly with respect to the applied tensile stresses. The stiffness of the fibers can be evaluated as the slope of the linear region of the stress–strain curve. Eventually, as the stretching continues, fibers begin to reach their loading capacity, and a nonlinear stress–strain response is observed. This is due to gradual rupture of the fibers. As the applied load increases, the remaining fibers will rupture leading to tissue failure as shown in Figure 2.11. The ultimate rupture stress and strain of human tissues were reported to be in the range 50–100 MPa, and 10%–15%, respectively (Fung 1993).

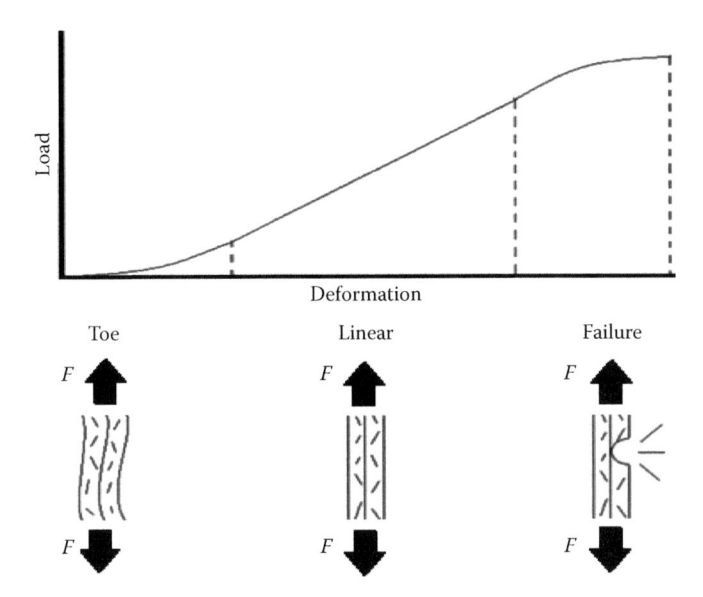

FIGURE 2.11   Collagen-fiber response to tensile loading.

The geometry of a ligament or a tendon affects its mechanical behavior (Whiting and Zernicke 2008). For instance, if two ligaments of the same cross section are subjected to uniaxial tensile loading, where one is twice the length of the other and both have the same cross section (same number of collagen fibers), the shorter ligament will exhibit twice the stiffness of the longer ligament, as illustrated in Figure 2.12a. However, the ultimate strength of both ligaments remains the same. The cross-sectional area also plays a significant role in defining the behavior of these tissues. For example, if two ligaments have different cross sections resulting in a different number of collagen fibers, and both were subjected to stretching, the ligament with the larger cross section will be stiffer than the ligament with a smaller cross section. However, both will experience similar elongation (Nordin and Frankel 2012), as shown in Figure 2.12b.

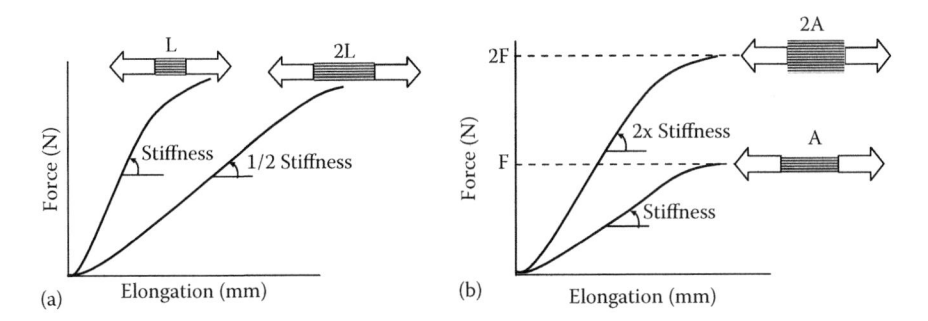

FIGURE 2.12   Effect of geometry on the mechanical response of connective tissues and skeletal muscles: (a) length effect and (b) cross-sectional area effect. (Adapted from Butler, D.L. et al., *Exerc. Sport Sci. Rev.*, 6, 125–181, 1978. With permission.)

One example for such a unique behavior was observed by stretching anterior cruciate ligament (ACL) of a rabbit. Kwan and Woo (1989) developed a structural model to evaluate this mechanical response. They presented a model of a single collagen fibril based on a bilinear assumption of the stress–strain relationship, expressed as follows:

$$\sigma = \begin{cases} E_s\varepsilon, & 0 < \varepsilon \le \varepsilon_s \\ E_e(\varepsilon - \varepsilon_s) + E_s\varepsilon_s, & \varepsilon_s < \varepsilon \le \varepsilon_u \end{cases} \tag{2.16}$$

where:

σ and ε are the tensile stress and strain, respectively
E_s is the initial stiffness
E_e is the elastic stiffness
ε_s is the strain in the uncrimped state of the fibril
ε_u is the ultimate strain

The model was then generalized to produce a more detailed mechanical behavior. The generalized form is expressed as follows:

$$\sigma = \begin{cases} E_s\varepsilon & 0 < \varepsilon \le \varepsilon_{s1} \\[2mm] [E_s + \gamma_1(E_e - E_s)]\varepsilon - (E_e - E_s)\gamma_1 E_{s1} & \varepsilon_{s1} < \varepsilon \le \varepsilon_{s2} \\[2mm] \vdots \\[2mm] \left[E_s + (E_e - E_s)\sum_{i=1}^{m-1}\gamma_i\right]\varepsilon - (E_e - E_s)\sum_{i=1}^{m-1}\gamma_i\varepsilon_{si} & \varepsilon_{s(m-1)} < \varepsilon \le \varepsilon_{sm} \\[2mm] E_e\varepsilon - (E_e - E_s)\sum_{i=1}^{m-1}\gamma_i\varepsilon_{si} & \varepsilon_{sm} < \varepsilon \le \varepsilon_{u1} \\[2mm] (1 - \beta_1)E_e\varepsilon - (E_e - E_s)\sum_{i=1}^{m}\gamma_i\varepsilon_{si} + \beta_1\varepsilon_{u1}E_e & \varepsilon_{u1} < \varepsilon \le \varepsilon_{u2} \\[2mm] \vdots \\[2mm] \beta_n E_e\varepsilon - (E_e - E_s)\sum_{i=1}^{m}\gamma_i\varepsilon_{si} + E_e\sum_{i=1}^{n-1}\beta_i\varepsilon_{ui} & \varepsilon_{u(n-1)} < \varepsilon < \varepsilon_{un} \end{cases} \tag{2.17}$$

where:

$\gamma_1, \gamma_2, \ldots\ldots, \gamma_m$ are the m group of collagen fibril lengths, which become fully extended at strain levels $\varepsilon_{s1}, \varepsilon_{s2}, \ldots\ldots, \varepsilon_{sm}$, respectively
$\beta_1, \beta_2, \ldots\ldots, \beta_n$ represent the n failure group of the extended fibrils that fail at strain levels $\varepsilon_{u1}, \varepsilon_{u2}, \ldots\ldots, \varepsilon_{un}$

This model successfully captured the mechanical behavior of the ACL.

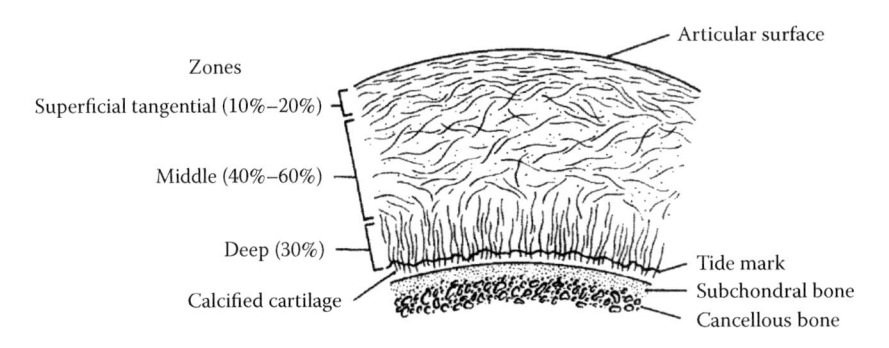

FIGURE 2.13 Random distribution of the fibrous structures in articular cartilage. (Reprinted from Mow, V.C. et al., *Biomaterials* 13, 67–97, 1992. With permission.)

Articular cartilage, another connective tissue, also consists of a fibrous structure. Type-II collagen is the main fibrous protein, which represents 90% of the collagen in the cartilage (Whiting and Zernicke 2008). In stretching, the mechanical response is, to some extent, similar to the ligaments and tendons; however, it does not develop the linear region in the force-deformation response as in stretching of ligaments and tendons. This behavior is related to the random distribution of collagen fibers, shown in Figure 2.13, which prevents the complete recruitment of these fibers under stretching, and also makes the indentation test the most common testing technique for this tissue.

As in other biological tissues, the musculoskeletal systems are also known for their time-dependent response. For example, they tend to show different mechanical behaviors under different strain rates, where a higher strain rate leads to a stiffer response (Vuskovic 2001).

Under tensile loading, it is also observed that if a tendon is subjected to a certain level of deformation, the force required to maintain the deformed length decreases with time. This phenomenon is known as stress relaxation, as shown in Figures 2.14 and 2.15. Similarly, if a tendon is subjected to a certain level of stretching, the load remains constant, and the tendon shows signs of creep. That is, strain increases with time under constant load. Similar creep behavior is observed in articular cartilage using the indentation test, as clearly demonstrated in Figure 2.16.

The viscoelastic response is attributed to the fact that tissues lose energy in the form of heat during deformation (Whiting and Zernicke 2008). Therefore, during loading–unloading cycles, the mechanical response shows retardation, forming what is known as *Hysteresis loop*, and the area within the loop represents the energy loss due to the viscous resistance, shown in Figure 2.17.

2.4.3 Internal Organs

Internal biological organs control life-sustaining functions of the human body. Organs such as the brain, kidney, lungs, liver, and spleen are commonly known as very soft tissues because of their relatively low stiffness properties. Over the past decade, there has been a growing interest in the mechanics of these tissues, prompted by the development

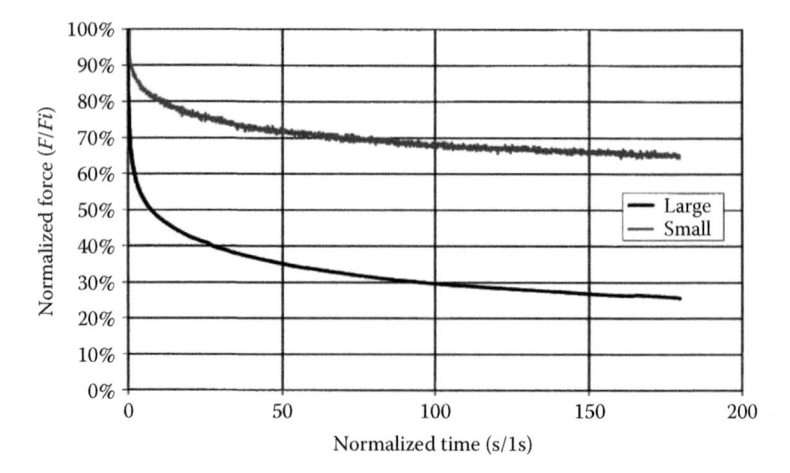

FIGURE 2.14   Stress relaxation of large and small human patellar tendons. (Reprinted from Atkinson, T.S. et al., *J Biomech*, 32, 907–914, 1999. With permission.)

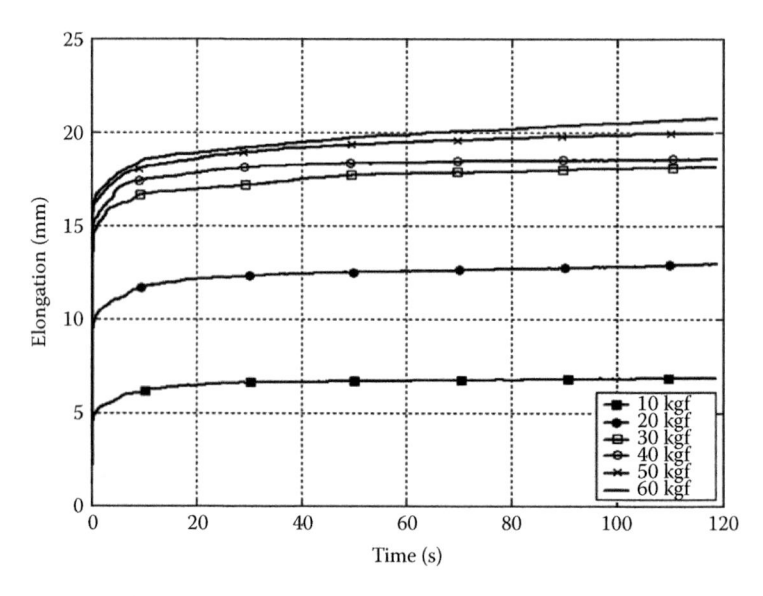

FIGURE 2.15   Creep response of muscular bulk tissue of 32-year-old healthy male under static loading of 10, 20, 30, 40, 50, and 60 kgf. (Reprinted from Arıtan, S. et al., *J. Biomech.*, 41, 2760–2765, 2008. With permission.)

in robotic surgeries such as the laparoscopic surgeries. This type of surgery includes minimally invasive techniques, such as robotic liver-resection surgery. In this surgery, special surgical tools are introduced into the abdomen through very small incisions (0.5–1.5 cm), which requires high eye–hand coordination from the surgeon. Therefore, surgical simulation systems are developed for training purposes. These systems are developed to provide, to some extent, realistic physical and visual responses

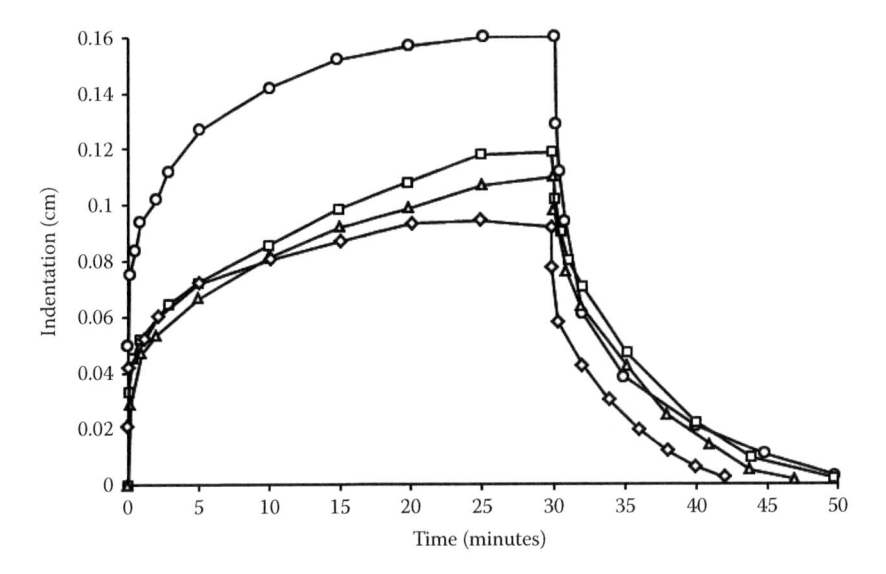

FIGURE 2.16 Creep of human cartilage in four areas under a load of 2.28 kgf. (Reprinted from Kempson, G.E. et al., *J. Biomech.*, 4, 239–250, 1971. With permission.)

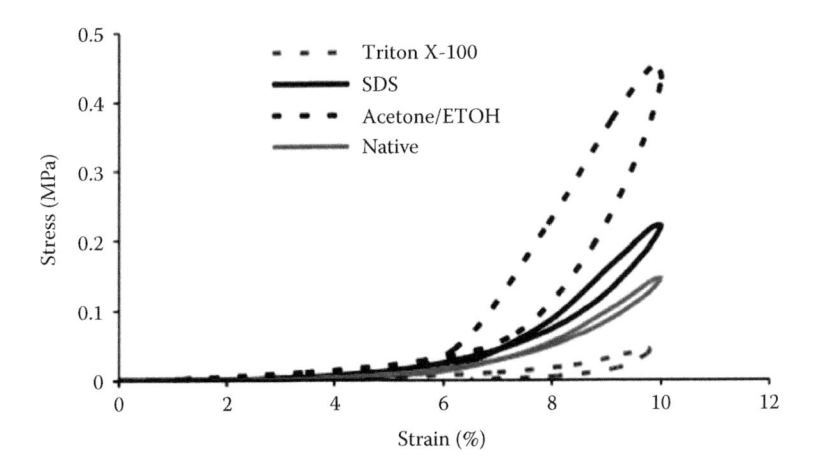

FIGURE 2.17 Hysteresis loop during the first cycle of loading of different types of tissue-engineered joint discs. (Reprinted from Lumpkins, S.B. et al., *Acta Biomaterialia*, 4, 808–816, 2008. With permission.)

of the simulated human organs. Thus, the systems essentially depend on models that allow real-time deformations such as in Cotin et al. (1999), Debunne et al. (2001), and Picinbono et al. (2003). These models are based on the mechanical behavior of internal biological organs. The mechanical response of these tissues is commonly characterized by observing their response under applied deformations. The following sections will highlight the response of the brain, liver, and spleen to tension and indentation.

2.4.3.1 Brain

The brain is subjected to many impact scenarios, such as sport and automobile accidents (O'Riordain et al. 2003, Rueda and Gilchrist 2009), resulting in traumatic injuries ranging from minor damage, such as concussion, to serious injuries, such as diffuse axonal injury (DAI). In the latter scenario, some cases can affect the function control of biological organs as the axon of the nerve cell is severely damaged. To study the response of the brain during these traumatic events, numerous research studies have developed numerical models to investigate brain dynamics during various loading conditions (Kleiven and Hardy 2002, Raul et al. 2006, Kleiven 2007, Ho and Kleiven 2009, Rueda and Gilchrist 2009, Zhang et al. 2011). These models are dependent on the mechanical properties of the biological tissues to present a relatively biofidelic prediction. In the field of neuroscience, the brain structure is divided into white and gray matters. The white matter consists of axons (myelinated nerve cells), and the gray matter consists of somas (cell bodies). The axons extend through the white matter to form its fibrous structure. Several studies have investigated the mechanical properties of brain tissues, as unimatter, using the tensile testing technique (Miller and Chinzei 2002, Velardi et al. 2006, Tamura et al. 2008). Rashid et al. (2014) evaluated the stiffness of brain tissues under stretching, as shown in Figure 2.18. Specifically, they evaluated the elastic moduli E_1, E_2, and E_3 at strain ranges of 0%–10%, 10%–20%, and 20%–30%, respectively, for porcine brain samples with different diameters (Rashid et al. 2014). Their results are summarized in Table 2.1.

Using scanning force microscopy techniques, the gray matter is 60% stiffer than the white matter (Christ et al. 2010). However, using magnetic resonance elastography, Feng et al. (2013) reported that white matter and gray matter have similar stiffness. These tangible differences might be attributed to the conditions of testing or the postmortem interval at which the samples were tested (Bordas and Balint 2015). Using indentation, a number of studies addressed the mechanical properties of the white and gray matter individually (van Dommelen et al. 2010, Budday et al. 2015). Although using different types of indenters (flat punch and spherical), it was reported that the white matter was stiffer and more viscous than the gray matter, as shown in Figure 2.19.

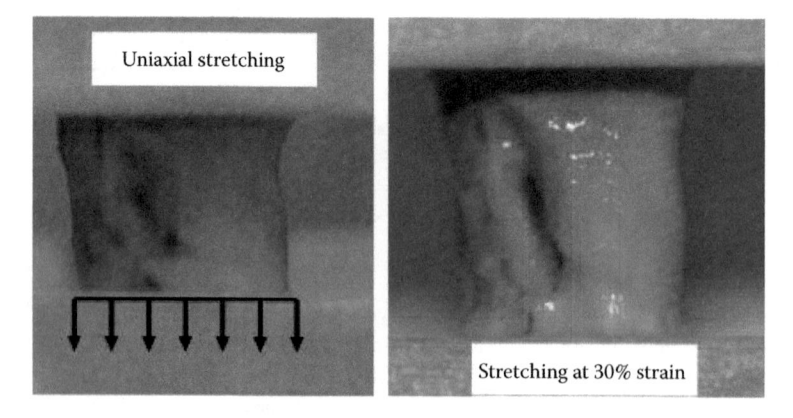

FIGURE 2.18 Brain sample of 10 mm length stretching from one side to achieve 30% strain. (Reprinted from Rashid, B. et al., *J. Mech. Behav. Biomed. Mater.*, 33, 43–54, 2014. With permission.)

TABLE 2.1 Elastic Moduli of the Brain Tissue Samples; These Data Represent the Average of the Tissues' Stiffness at Each Rate

Strain Rate (1/s)	E_1	E_2	E_3
Strain range	0%–10%	10%–20%	20%–30%
30	8.12	19.2	29.46
60	10.86	28	41.05
90	16.08	35.6	60.73

Source: Rashid, B. et al., *J. Mech. Behav. Biomed. Mater.*, 33, 43–54, 2014.

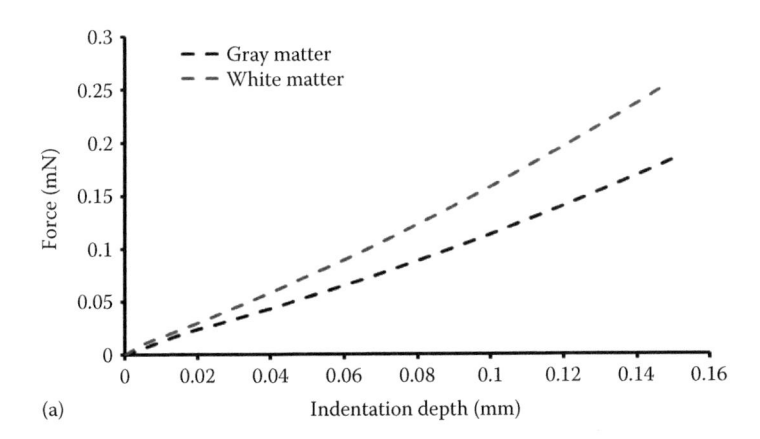

(a)

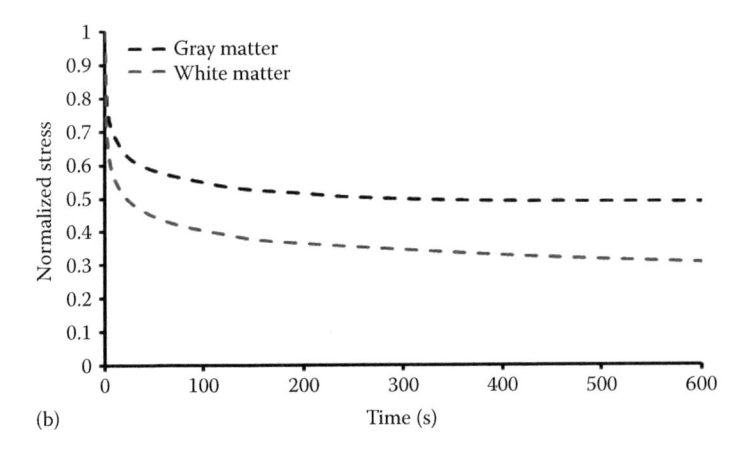

(b)

FIGURE 2.19 (a) Force–displacement data of indenting the white matter and gray matter of brain tissues (Data from van Dommelen, J.A.W. et al., *J. Mech. Behav. Biomed. Mater.*, 3, 158–166, 2010.) (b) relaxation test results reflect that the white matter is more viscous than the gray matter. (Data from Budday, S. et al., *J. Mech. Behav. Biomed. Mater.*, 46, 318–330, 2015.)

Brain viscoelasticity is commonly studied using tensile stretching. Normally, brain tissues exhibit viscoelastic behavior if they are subjected to different stretching rates. Rashid et al. (2014) applied 30/s, 60/s, and 90/s strain velocities on fresh brain samples of 6-month-old pigs. Their results indicated a stiffer stress–stretch relationship as loading rate increased, as illustrated in Table 2.1, which reflected higher elastic moduli at higher stretching velocities. In addition, the viscoelastic response of soft tissues is affected by the structural constituents of these tissues at the microlevel (Sack et al. 2009). Viscoelastic properties of brain tissues can be an indication to aging effects. As the brain ages, it experiences degradation of oligodendrocytes and neurons (Morrison and Hof 1997). Furthermore, change in brain tissue viscoelasticity has been observed in many diseases such as Alzheimer's (Murphy et al. 2016) and multiple sclerosis (Wuerfel et al. 2010, Streitberger et al. 2012).

2.4.3.2 Liver and Spleen

Investigating the mechanics of abdominal tissues such as the liver and spleen assists in understanding the mechanisms of injuries and a deeper understanding of their deformation for different medical interventions such as radiotherapy and surgery. For example, in automobile accidents, the abdominal region of the vehicle's occupants is subjected to a variety of impact scenarios. These scenarios can lead to liver laceration, which is an injury that occurs due to severe abdominal trauma and causes the tear of liver capsule and parenchyma. This is attributed to the relatively large amount of blood in the liver, which generates excessive pressure during an impact event and results in tension stresses on its surface (Brunon et al. 2010).

As in other internal organs, the liver consists of parenchymal tissue that is responsible for their biological functions. The parenchyma is covered with a dense fibrous tissue known as the capsule. The behavior of liver parenchyma and capsule tissues under tensile loads was reported in Brunon et al. (2010), shown in Figure 2.20. Under stretching, the liver first exhibits the low stiffness region (toe region) due to the extension of randomly arranged collagen fibers. Afterward, a linear region is observed as the collagen fibers are fully extended. At failure, the capsule fails first, resulting in a sudden drop of load until the parenchyma starts to resist the applied loads, which can be observed as a second strain hardening. Eventually, the parenchyma fails and releases the stored energy due to stretching.

Both the liver parenchyma and capsule tend to exhibit nonlinear responses when subjected to uniaxial tensile loads. However, each has its own distinct behavior. Liver parenchyma was reported to experience a failure stress of 61 kPa when deformation was applied at a strain rate of $10\ s^{-1}$, and the failure strain was 24% (Kemper et al. 2010). On the other hand, the liver capsule showed harder behavior at a lower strain rate ($0.5\ s^{-1}$), and failed at 9.2 MPa stress and 35.6% strain (Hollenstein et al. 2006). At failure, biological tissues experience high stresses and exhibit lower deformation when high strain rates are applied. By comparing responses of the liver parenchyma and capsule at failure, the liver capsule showed higher strength at lower strain rates, whereas the liver parenchyma showed lower strength at higher strain rates, which indicates that liver parenchyma is softer than its capsule.

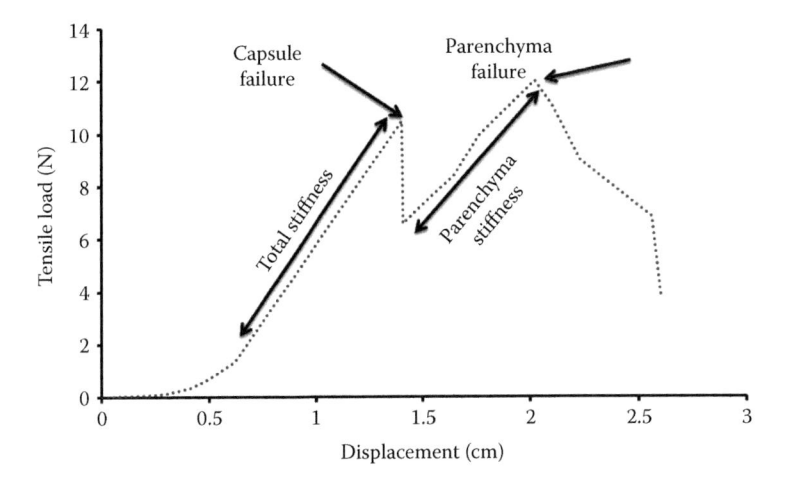

FIGURE 2.20 Tensile load–displacement of liver tissue. Liver parenchyma is encapsulated inside the liver capsule. (Reprinted from Brunon, A.K. et al., *J. Biomech.*, 43, 2221–2227, 2010. With permission.)

The spleen is also susceptible to injury, such as blunt abdominal trauma (Dupuy 1995, Peitzman et al. 2001), resulting in increased mortality rates in patients over 55-year-old (Harbrecht et al. 2001). These splenic injuries can be as severe as ruptures or lacerations that involve the spleen capsule and parenchyma (Peitzman et al. 2001). Blunt trauma is very common in automobile accidents. Currently, crush test dummies, known as anthropomorphic test devices (ATDs), are used to improve the safety systems of modern cars; however, these dummies are not designed to predict the injury levels experienced by occupants during accidents. Therefore, safety engineers depend mainly on numerical models to predict the probabilities of having abdominal injuries such as splenic injuries and associated tissue-tolerance levels (Kemper et al. 2012). Studying the mechanics of splenic tissues provides the necessary biomechanical data required to create material models that mimic spleen tissues and predict their behavior under normal, pathological, and traumatic conditions.

The mechanical response of spleen tissues subjected to tensile loading is characterized by the common S-shaped stress–strain relationship. The response of spleen tissue to tensile loads is generally softer than liver. Figure 2.21 shows liver parenchyma failed at stress and strain of 40.21 kPa and 34%, respectively, when stretching was applied with a strain rate of 0.01 s^{-1}. On the other hand, spleen parenchyma failed at 16.5 kPa and 26% stress and strain using the same strain rate (Kemper et al. 2012).

The viscoelastic behavior of liver and spleen tissues was reported in Kemper et al. (2010) and (2012), where 51 liver samples and 41 spleen samples were tested. These samples were extracted from 7 human livers and 14 human spleens, and were subjected to uniaxial tensile tests at 4 different strain rates. The samples experienced higher failure stresses and lower failure strains as the strain rates increased, as presented in Table 2.2.

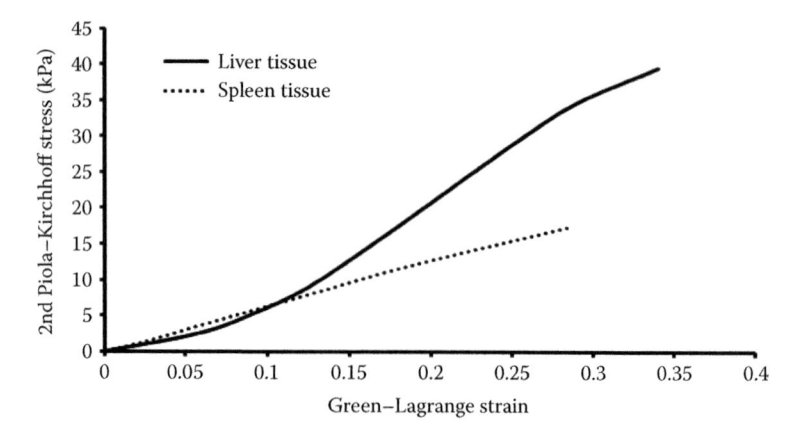

FIGURE 2.21 Stretching response of liver and spleen tissues. (Data from Kemper, A.R. et al., *Annals of Advances in Automotive Medicine/Annual Scientific Conference*, 54, 15–26, 2010; Kemper, A.R. et al., *J. Biomech.*, 45, 348–355, 2012.)

TABLE 2.2 Human Liver Parenchyma Sample Stretching and Their Corresponding Failure Stresses and Failure Strain

		Liver		Spleen	
	Strain Rate (s^{-1})	Failure Stress (kPa)	Failure Strain %	Failure Stress (kPa)	Failure Strain %
Rate 1	0.01	40.21	34	16.5	26
Rate 2	0.1	46.79	32	23.9	21
Rate 3	1	52.61	30	31.5	19
Rate 4	10	61.02	24	33.8	18

Source: Kemper, A.R. et al., *Annals of Advances in Automotive Medicine/Annual Scientific Conference*, 54, 15–26, 2010; Kemper, A.R. et al., *J. Biomech.*, 45, 348–355, 2012.

Biological tissues behave differently when tested *in vivo* and *ex vivo*. In general, the tensile test is a reliable technique for studying the mechanics of biological tissues; however, it has been commonly used only in studying extracted *ex vivo* biological tissues. Investigating the mechanics of organ tissue within its physiological environment (*in vivo*) poses great challenges. On the contrary, the indentation technique can be used to investigate the mechanical behavior of biological tissues *in vivo* such as the one developed by Carter et al. (2001). They used an indentation device to measure stiffness of the liver right loop of six patients. The average stiffness of these samples was 0.27 MPa with a standard deviation of around 30%. However, they incurred a high standard deviation, caused by one particular liver that experienced cholestasis. The stiffness of the diseased liver was 0.74 MPa, nearly three times the average stiffness.

Brown et al. (2003) investigated the difference between *in vivo* and *ex vivo* samples by applying deformations at a rate of 5 mm/s through 5 cycles on the same porcine liver, first *in vivo*, then *ex vivo*. The stiffness was generally the same between each liver condition using different rates. However, response was relatively different at each loading cycle. On the other hand, when the same liver was tested *ex vivo*, only the first loading cycle was relatively different, and the liver responded similarly in the subsequent four cycles.

2.5 Conclusion

- Understanding the mechanics of soft biological tissues plays a fundamental role in the field of biomechanics. It improves the safety standards and designs in the automotive and sports industries. It can be used to investigate the mechanics of diseases for diagnostic and therapeutic purposes.
- The mechanical properties of biological tissues have been investigated using numerous techniques. The tensile test and indentation test are among the most common techniques. Both techniques are direct and are based on applying deformations and observing the forces experienced by the tissues.
- The tensile test is commonly applied to *ex vivo* samples of soft tissues. Generally, it is conducted by applying uniaxial or biaxial loading. Biaxial testing applies more boundary conditions than the uniaxial test, which results in considerable differences in the materials' behavior between these methods.
- Indentation is based on contact mechanics and can be implemented on soft tissues, *ex vivo* and *in vivo*. The material response to loading depends significantly on the types of indenter used in the test. Soft tissues tend to exhibit higher contact stiffness when indented with flat or spherical indenters than sharp indenter.
- Tissues from the musculoskeletal systems, such as ligaments and tendons, are fibrous and composed broadly of collagen fibers, which significantly contribute to high strength of these tissues. This type of tissue is normally stretched to investigate their mechanics based on the fiber's arrangement inside these tissues.
- The tissues from internal organs such as the brain, spleen, and liver exhibit lower stiffness behavior when compared with ligaments and tendons.
- Brain tissues can generally be divided into two types: (1) gray and (2) white matter. Although the tensile tests can give an overall idea about the mechanical behavior of brain tissues, indentation is used to test the gray and white matter individually to characterize their mechanics separately.
- The liver and spleen are composed of two types of tissues: (1) parenchyma and (2) capsule. The capsule lines the parenchyma and is generally stiffer. In general, liver tissues are stiffer than the spleen tissue.
- The biological tissues are known for their viscoelastic response, or time-dependent behavior, including creep, relaxation, and hysteresis.

References

Arıtan, S., S. O. Oyadiji, and R. M. Bartlett. 2008. A mechanical model representation of the *in vivo* creep behaviour of muscular bulk tissue. *Journal of Biomechanics* 41 (12): 2760–2765.

Atkins, A. G., and D. Tabor. 1965. Plastic indentation in metals with cones. *Journal of the Mechanics and Physics of Solids* 13 (3): 149–164.

Atkinson, T. S., B. J. Ewers, and R. C. Haut. 1999. The tensile and stress relaxation responses of human patellar tendon varies with specimen cross-sectional area. *Journal of Biomechanics* 32 (9): 907–914.

Blake, A. (Ed.). 1985. *Handbook of Mechanics, Materials, and Structures.* Wiley Series in Mechanical Engineering Practice. New York: Wiley.

Bordas, S. P. A, and D. S. Balint. 2015. *Advances in Applied Mechanics.* Volume 48.

Brown, J. D., J. Rosen, M. N. Sinanan, and B. Hannaford. 2003. In-vivo and postmortem compressive properties of porcine abdominal organs. In *Medical Image Computing and Computer-Assisted Intervention—MICCAI 2003*, Randy E. Ellis and Terry M. Peters. (Eds.), 2878: 238–245. Berlin, Germany: Springer.

Brunon, A., K. Bruyère-Garnier, and M. Coret. 2010. Mechanical characterization of liver capsule through uniaxial quasi-static tensile tests until failure. *Journal of Biomechanics* 43 (11): 2221–2227.

Budday, S., R. Nay, R. de Rooij, P. Steinmann, T. Wyrobek, T. C. Ovaert, and E. Kuhl. 2015. Mechanical properties of gray and white matter brain tissue by indentation. *Journal of the Mechanical Behavior of Biomedical Materials* 46: 318–330.

Butler, D. L., E. S. Grood, F. R. Noyes, and R. F. Zernicke. 1978. Biomechanics of ligaments and tendons. *Exercise and Sport Sciences Reviews* 6: 125–181.

Cao, Y. P., M. Dao, and J. Lu. 2007. A precise correcting method for the study of the superhard material using nanoindentation tests. *Journal of Materials Research* 22 (5): 1255–1264.

Carter, F. J., T. G. Frank, P. J. Davies, D. McLean, and A. Cuschieri. 2001. Measurements and modelling of the compliance of human and porcine organs. *Medical Image Analysis* 5 (4): 231–236.

Chaudhri, M. M. 2001. A note on a common mistake in the analysis of nanoindentation data. *Journal of Materials Research* 16 (2): 336–339.

Christ, A. F., K. Franze, H. Gautier, P. Moshayedi, J. Fawcett, R. J. Franklin, R. T. Karadottir, and J. Guck. 2010. Mechanical difference between white and gray matter in the rat cerebellum measured by scanning force microscopy. *Journal of Biomechanics* 43 (15): 2986–2992.

Cotin, S., H. Delingette, and N. Ayache. 1999. Real-time elastic deformations of soft tissues for surgery simulation. *IEEE Transactions on Visualization and Computer Graphics* 5 (1): 62–73.

Dale, W. C., E. Baer, A. Keller, and R. R. Kohn. 1972. On the ultrastructure of mammalian tendon. *Experientia* 28 (11): 1293–1295.

Debunne, G., M. Desbrun, M. P. Cani, and A. H. Barr. 2001. Dynamic real-time deformations using space & time adaptive sampling. In *Proceedings of the 28th Annual Conference on Computer Graphics and Interactive Techniques* (pp. 31–36). ACM.

Deplano, V., M. Boufi, O. Boiron, C. Guivier-Curien, and Y. Alimi. 2016. Biaxial tensile tests of the porcine ascending aorta. *Journal of Biomechanics* 49 (10): 2031–2037.

van Dommelen, J. A. W., T. P. J. van der Sande, M. Hrapko, and G. W. M. Peters. 2010. Mechanical properties of brain tissue by indentation: Interregional variation. *Journal of the Mechanical Behavior of Biomedical Materials* 3 (2): 158–166.

Duncan, B. C., A. S. Maxwell, L. E. Crocker, and R. Hunt. 1999. Verification of hyperelastic test methods. PAJ1 Report 17.

Dupuy, D. E., V. Raptopoulos, and M. P. Fink. 1995. Current concepts in splenic trauma. *Journal of Intensive Care Medicine* 10 (2): 76–90.

Ebenstein, D. M., and L. A. Pruitt. 2004. Nanoindentation of soft hydrated materials for application to vascular tissues. *Journal of Biomedical Materials Research* 69A (2): 222–232.

Feng, Y., E. H. Clayton, Y. Chang, R. J. Okamoto, and P. V. Bayly. 2013. Viscoelastic properties of the ferret brain measured in vivo at multiple frequencies by magnetic resonance elastography. *Journal of Biomechanics* 46 (5): 863–870.

Fischer-Cripps, A. C. 2001. Use of combined elastic modulus in the analysis of depth-sensing indentation data. *Journal of Materials Research* 16 (11): 3050–3052.

Fischer-Cripps, A. C. 2003. Use of combined elastic modulus in depth-sensing indentation with a conical indenter. *Journal of Materials Research* 18 (5): 1043–1045.

Fischer-Cripps, A. C. 2007. *Introduction to Contact Mechanics.* 2nd ed. Mechanical Engineering Series. New York: Springer.

Fischer-Cripps, A. C. 2011. *Nanoindentation.* Mechanical Engineering Series. New York: Springer.

Fratzl, P., and R. Weinkamer. 2007. Nature's hierarchical materials. *Progress in Materials Science* 52 (8): 1263–1334.

Fung, Y. C. 1993. *Biomechanics Mechanical Properties of Living Tissues.* New York: Springer.

Gouldstone, A., N. Chollacoop, M. Dao, J. Li, A. Minor, and Y. Shen. 2007. Indentation across size scales and disciplines: Recent developments in experimentation and modeling. *Acta Materialia* 55 (12): 4015–4039.

Gregory, D. E., and J. P. Callaghan. 2011. A comparison of uniaxial and biaxial mechanical properties of the annulus fibrosus: A porcine model. *Journal of Biomechanical Engineering* 133 (2): 024503.

Harbrecht, B. G., A. B. Peitzman, L. Rivera, B. Heil, M. Croce, J. A. Morris Jr., B. L. Enderson. 2001. Contribution of age and gender to outcome of blunt splenic injury in adults: Multicenter study of the eastern association for the surgery of trauma. *The Journal of Trauma* 51 (5): 887–895.

Hertz, H. 1881. On the contact of elastic solids. In *Translated and Reprinted in English in Hertz's Miscellaneous Papers*, Macmillan, London, 1896, Ch5, 156–171.

Ho, J., and S. Kleiven. 2009. Can sulci protect the brain from traumatic injury? *Journal of Biomechanics* 42 (13): 2074–2080.

Hollenstein, M., A. Nava, D. Valtorta, J. G. Snedeker, and E. Mazza. 2006. Mechanical characterization of the liver capsule and parenchyma. In *Biomedical Simulation*, M. Harders and G. Székely. (Eds.). 150–58. Lecture Notes in Computer Science 4072. Berlin, Germany: Springer.

Holzapfel, G. A. 2004. Computational biomechanics of soft biological tissue. In *Encyclopedia of Computational Mechanics*, E. Stein, R. de Borst, and T. J. R. Hughes. (Eds.). Chichester, UK: John Wiley & Sons.

Jacobs, N. T., D. H. Cortes, E. J. Vresilovic, and D. M. Elliott. 2013. Biaxial tension of fibrous tissue: Using finite element methods to address experimental challenges arising from boundary conditions and anisotropy. *Journal of Biomechanical Engineering* 135 (2): 021004.

Kemper, A. R., A. C. Santago, J. D. Stitzel, J. L. Sparks, and S. M. Duma. 2010. Biomechanical response of human liver in tensile loading. *Annals of Advances in Automotive Medicine/Annual Scientific Conference*. Association for the Advancement of Automotive Medicine. Scientific Conference 54: 15–26.

Kemper, A. R., A. C. Santago, J. D. Stitzel, J. L. Sparks, and S. M. Duma. 2012. Biomechanical response of human spleen in tensile loading. *Journal of Biomechanics* 45 (2): 348–355.

Kempson, G. E., M. A. R. Freeman, and S. A. V. Swanson. 1971. The determination of a creep modulus for articular cartilage from indentation tests on the human femoral head. *Journal of Biomechanics* 4 (4): 239–250.

Kleiven, S. 2007. Predictors for traumatic brain injuries evaluated through accident reconstructions. *Stapp Car Crash Journal* 51: 81–114.

Kleiven, S., and W. N. Hardy. 2002. Correlation of an FE model of the human head with local brain motion: Consequences for injury prediction. *Stapp Car Crash Journal* 46: 123–144.

Kwan, M. K., and S. L. Woo. 1989. A structural model to describe the nonlinear stress-strain behavior for parallel-fibered collagenous tissues. *Journal of Biomechanical Engineering* 111 (4): 361–363.

Liu, K., M. R. VanLandingham, and T. C. Ovaert. 2009. Mechanical characterization of soft viscoelastic gels via indentation and optimization-based inverse finite element analysis. *Journal of the Mechanical Behavior of Biomedical Materials* 2 (4): 355–363.

Lumpkins, S. B., N. Pierre, and P. S. McFetridge. 2008. A mechanical evaluation of three decellularization methods in the design of a xenogeneic scaffold for tissue engineering the temporomandibular joint disc. *Acta Biomaterialia* 4 (4): 808–816.

Lynch, H. A. 2003. Effect of fiber orientation and strain rate on the nonlinear uniaxial tensile material properties of tendon. *Journal of Biomechanical Engineering* 125 (5): 726.

Miller, K., and K. Chinzei. 2002. Mechanical properties of brain tissue in tension. *Journal of Biomechanics* 35 (4): 483–490. doi:10.1016/S0021-9290(01)00234-2.

Morrison, J. H., and P. R. Hof. 1997. Life and death of neurons in the aging brain. *Science* 278 (5337): 412–419.

Mow, V. C., A. Ratcliffe, and A. R. Poole. 1992. Cartilage and diarthrodial joints as paradigms for hierarchical materials and structures. *Biomaterials* 13 (2): 67–97.

Mucsi, A. 2013. Effect of gripping system on the measured upper yield strength estimated by tensile tests. *Measurement* 46 (5): 1663–1670.

Murphy, M. C., D. T. Jones, C. R. Jack, K. J. Glaser, M. L. Senjem, A. Manduca, J. P. Felmlee, R. E. Carter, R. L. Ehman, and J. Huston. 2016. Regional brain stiffness changes across the Alzheimer's disease spectrum. *NeuroImage: Clinical* 10: 283–290.

Nordin, M., and V. H. Frankel. (Eds.) 2012. *Basic Biomechanics of the Musculoskeletal System*. 4th ed. Philadelphia, PA: Wolters Kluwer Health/Lippincott Williams & Wilkins.

O'Riordain, K., P. M. Thomas, J. P. Phillips, and M. D. Gilchrist. 2003. Reconstruction of real world head injury accidents resulting from falls using multibody dynamics. *Clinical Biomechanics* 18 (7): 590–600.

Oyen, M. L. (Ed.). 2011. *Handbook of Nanoindentation with Biological Applications*. Singapore: Pan Stanford Publishing.

Peitzman, A. B., H. R. Ford, B. G. Harbrecht, D. A. Potoka, and R. N. Townsend. 2001. Injury to the spleen. *Current Problems in Surgery* 38 (12): 932–1008.

Picinbono, G., H. Delingette, and N. Ayache. 2003. Non-linear anisotropic elasticity for real-time surgery simulation. *Graphical Models* 65 (5): 305–321.

Rashid, B., M. Destrade, and M. D. Gilchrist. 2014. Mechanical characterization of brain tissue in tension at dynamic strain rates. *Journal of the Mechanical Behavior of Biomedical Materials* 33: 43–54.

Raul, J. S., D. Baumgartner, R. Willinger, and B. Ludes. 2006. Finite element modelling of human head injuries caused by a fall. *International Journal of Legal Medicine* 120 (4): 212–218.

Rueda, M. A. F., and M. D. Gilchrist. 2009. Comparative multibody dynamics analysis of falls from playground climbing frames. *Forensic Science International* 191 (1–3): 52–57.

Sack, I., B. Beierbach, J. Wuerfel, D. Klatt, U. Hamhaber, S. Papazoglou, P. Martus, and J. Braun. 2009. The impact of aging and gender on brain viscoelasticity. *NeuroImage* 46 (3): 652–657.

Sacks, M. S. 2000. Biaxial mechanical evaluation of planar biological materials. *Journal of Elasticity* 61 (1–3): 199–246.

Sneddon, I. N. 1946. Boussinesq's problem for a flat-ended cylinder. *Mathematical Proceedings of the Cambridge Philosophical Society* 42 (1): 29.

Sneddon, I. N. 1948. Boussinesq's problem for a rigid cone. *Mathematical Proceedings of the Cambridge Philosophical Society* 44 (4): 492. doi:10.1017/S0305004100024518.

Sneddon, I. N. 1965. The relation between load and penetration in the axisymmetric Boussinesq problem for a punch of arbitrary profile. *International Journal of Engineering Science* 3 (1): 47–57.

Streitberger, K., I. Sack, D. Krefting, C. Pfüller, J. Braun, F. Paul, and J. Wuerfel. 2012. Brain viscoelasticity alteration in chronic-progressive multiple sclerosis. *PLoS One* 7 (1): e29888.

Tamura, A., S. Hayashi, K. Nagayama, and T. Matsumoto. 2008. Mechanical characterization of brain tissue in high-rate extension. *Journal of Biomechanical Science and Engineering* 3 (2): 263–274.

Timoshenko, S., and J. N. Goodier. 1951. *Theory of Elasticity*. 2nd ed. New York: McGraw-Hill.

Tong, K. J., and D. M. Ebenstein. 2015. Comparison of spherical and flat tips for indentation of hydrogels. *JOM* 67 (4): 713–719.

Velardi, F., F. Fraternali, and M. Angelillo. 2006. Anisotropic constitutive equations and experimental tensile behavior of brain tissue. *Biomechanics and Modeling in Mechanobiology* 5 (1): 53–61.

Vuskovic, V. 2001. Device for in-vivo measurement of mechanical properties of internal human soft tissues. Zurich, Switzerland. (Doctoral dissertation).

Whiting, W. C., and R. F. Zernicke. 2008. *Biomechanics of Musculoskeletal Injury.* 2nd ed. Champaign, IL: Human Kinetics.

Wuerfel, J., F. Paul, B. Beierbach, U. Hamhaber, D. Klatt, S. Papazoglou, F. Zipp, P. Martus, J. Braun, and I. Sack. 2010. MR-elastography reveals degradation of tissue integrity in multiple sclerosis. *NeuroImage* 49 (3): 2520–2525.

Zemanek, M., J. Bursa, and M. Detak. 2009. Biaxial tension tests with soft tissues of arterial wall. *Engineering Mechanics* 16: 3–11.

Zhang, J., N. Yoganandan, F. A. Pintar, Y. Guan, B. Shender, G. Paskoff, and P. Laud. 2011. Effects of tissue preservation temperature on high strain-rate material properties of brain. *Journal of Biomechanics* 44 (3): 391–396.

3

Magnetic Resonance Elastography

Deirdre M. McGrath

Magnetic resonance elastography (MRE) is a recently developed technology that uses magnetic resonance imaging (MRI) to measure the biomechanical properties of biological tissues, such as elasticity and viscosity. Due to its sensitivity to pathology-driven alterations in tissue biomechanics, MRE is a powerful diagnostic tool for detecting and staging disease, including fibrosis, cancer, and inflammation. The technology is undergoing rapid development for application to multiple organ sites in the body, and is already widely adopted in clinical practice for diagnosing hepatic fibrosis. This chapter will provide an overview of the background, methodology, and clinical applications of MRE.

3.1 Overview of Elasticity Imaging

MRE methods fall within the broader general classification of elastography or elasticity imaging, which includes non-MRI-based methods. Elasticity imaging consists of three steps: (1) generation of a mechanical stress within the biological tissue, which could be the result of an applied external mechanical force, or an internal endogenous mechanism (e.g., breathing, heart motion, or blood vessel pulsation); (2) observation of the strain response (e.g., displacement field or velocity) through an imaging method such as MRI, ultrasound, or optical imaging; and (3) application of an inversion algorithm to calculate biomechanical properties from the observed stress–strain dynamics (Figure 3.1).

When a linearly elastic isotropic Hookean material-constitutive model of the biological tissue is employed, the typical parameters measured with elastography include the shear modulus (G) and Young's modulus (E), whereas for a viscoelastic material-constitutive model the complex shear modulus (G^*) is usually measured. Soft tissues typically have material properties that are intermediate between those of solids and fluids; and hence, mainly behave as viscoelastic materials. Near-incompressibility is another common feature of soft tissues, and hence Poisson's ratio (v) is typically in the range of 0.49–0.499, which is very close to its value for fluid (0.5). Near-incompressibility also implies that, while shear and Young's modulus values may differ between tissue types by several orders of magnitude, the bulk modulus

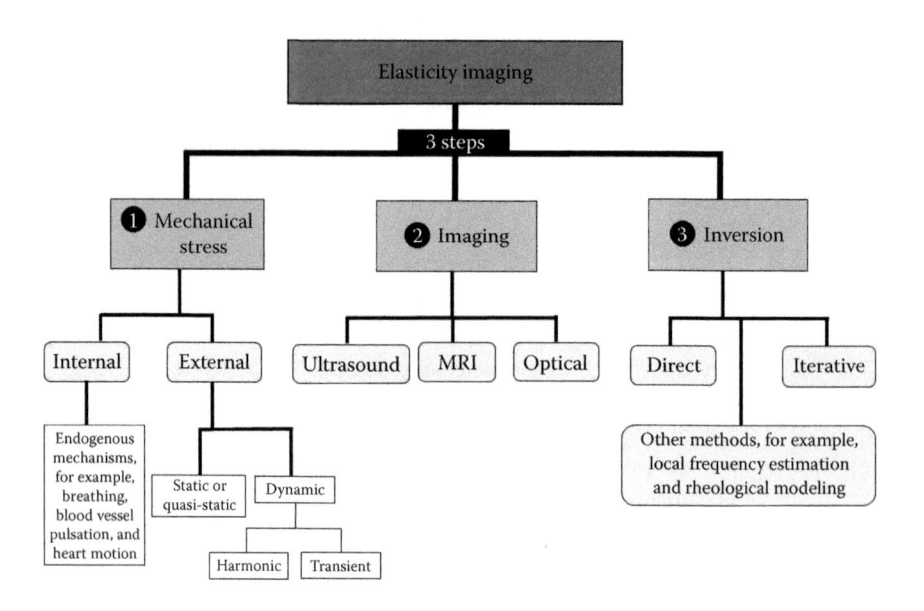

FIGURE 3.1   Flowchart of elasticity imaging methodological steps and subcategories.

(K) values typically differ from that of water by less than 15% (Goss et al. 1978). The density of soft tissue (ρ) is usually estimated as that of water (1000 kg/m^3). While some tissue types can be approximated as a material with isotropic Hookean material properties, most tissues have anisotropic, non-Hookean, and viscoelastic properties. Furthermore, some tissues are observed to have poroelastic material properties, that is, those of a solid matrix with fluid-filled pores.

3.2 Categories of Magnetic Resonance Elastography

There are two main classes of MRE: (1) dynamic MRE, which involves the delivery of mechanical waves (typically in the frequency range of 10–1000 Hz), which can be observed in the time harmonic steady-state, or in a transient state through various steps in their propagation; and (2) static or quasi-static MRE, in which a compressive force is applied to the tissue volume and the resultant strain field is measured. Initial MRE experiments with static loading began in 1989 (Sarvazyan et al. 1995, 2011), whereas dynamic MRE with delivery of mechanical waves was first demonstrated by Muthupillai et al. (1995).

In static elastography, a single compression is applied, and the displacement field (or strain field) of the whole tissue volume is measured in one step (Osman 2003). By contrast, quasi-static methods apply cyclically repeated compressions and collect different portions of the imaging data during each compression cycle. However, the frequency of the compression cycle is kept low (e.g., 1 Hz) to create conditions of approximate static stress. For *in vivo* applications, because of the technical difficulty of delivering a compressive force to sites deep within the body, static, and quasi-static methods are generally limited to more superficial tissues or easily accessible sites. For instance, *in vivo* quasi-static MRE techniques have been developed for breast (Plewes et al. 2000, Samani et al. 2001). However, most applications to date have been for *ex vivo* tissues (Chenevert et al. 1998, Samani et al. 2003, Hardy et al. 2005, McGrath et al. 2012).

Dynamic MRE is the more suitable methodology for *in vivo* investigations, as the mechanical waves can be delivered to sites deep within the body. Although there are two classes of dynamic MRE (harmonic and transient), comparatively little work has been carried out with transient MRE (McCracken et al. 2005, Souchon et al. 2008), and most dynamic MRE studies involve observation of the harmonic steady-state wave pattern. The frequencies employed for dynamic MRE are more commonly in the range of 10–100 Hz, as the wavelengths for these frequencies are in a measureable range of millimeters and centimeters, and higher frequencies are rapidly attenuated in viscoelastic tissue. Some *in vitro* dynamic MRE methods have also been developed (Othman et al. 2005), but drawbacks include the possibility of reflections from the tissue sample edges or container boundaries, which may result in interference patterns or standing waves, which could confound accurate inversion calculation of the biomechanical properties. Static and quasi-static methods avoid this problem by maintaining a state of static stress (or approximate static stress) during imaging.

Furthermore, other MRI-based methodologies have been developed for measuring strain distributions resulting from endogenous mechanisms such as breathing, heart motion, cerebrospinal fluid pulsation, or intravascular blood flow. These include motion-encoded phase-contrast imaging (O'Donnell 1985, Weaver et al. 2012, Hirsch et al. 2013), and saturation or spin tagging (Zerhouni et al. 1988, Axel and Dougherty 1989).

3.3 Comparison of Magnetic Resonance Elastography with Alternative Methods

3.3.1 Conventional Medical Imaging

MRE provides information on tissue biomechanics that cannot be derived from other conventional morphological (e.g., computed tomography, or CT) or physiological (functional)-imaging methods (e.g., nuclear medicine imaging). Furthermore, as the mechanical properties of biological tissue can vary enormously between different pathologic states and various types of healthy tissue, that is, by up to five orders of magnitude, elastography has a superior sensitivity to diseases as compared with conventional imaging methods (Mariappan et al. 2010) (Figure 3.2). In fact, no other physical

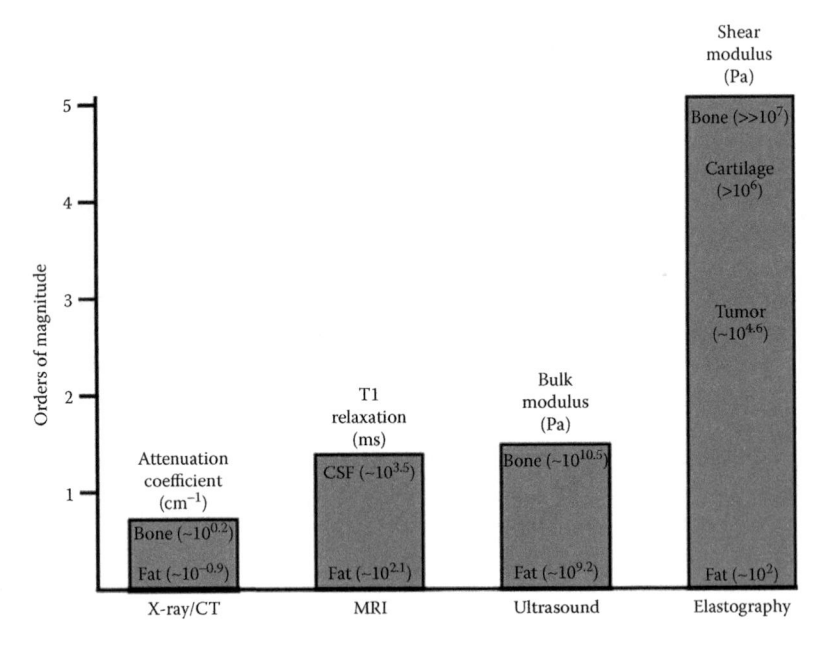

FIGURE 3.2 Imaging modality contrast mechanisms. Examples of different imaging modalities and the spectrum of contrast mechanisms utilized by them are shown. The shear modulus has the largest variation with variations over five orders of magnitude among various physiological states of normal and pathologic tissues. (Reprinted from Mariappan, Y. K. et al., *Clin. Anat.*, 23, 497–511, 2010. With permission.)

measurement of tissue properties changes as much as elasticity for different states of physiology and pathology (Manduca et al. 2001).

3.3.2 Manual Palpation

Traditionally, manual palpation has been employed by physicians for diagnosis through the detection of alterations in the mechanical stiffness of tissue, for example, for breast cancer. MRE can be considered a form of *virtual palpation* but holds important advantages over manual palpation: (1) MRE allows noninvasive measurement at sites deep within the body that are inaccessible to palpation; (2) MRE provides quantitative measures of the biomechanical properties, as opposed to the subjective qualitative assessment of the physician; (3) MRE provides a three-dimensional (3D) visualization of tissue biomechanics at high spatial resolution, as opposed to an average assessment for the bulk tissue volume; and (4) for diseases such as cancer, early detection greatly improves the treatment outcome, and palpation cannot provide the same sensitivity as MRE to such early stage changes.

3.3.3 Alternative Elasticity Imaging Methods

MRE also has advantages over alternative elasticity imaging methods, such as optical coherence tomography elastography (van Soest et al. 2007), tissue Doppler optical coherence elastography (Wang et al. 2006), ultrasound transient elastography (Sandrin et al. 2003), ultrasound shear-wave elasticity imaging (Sarvazyan et al. 1998), and mechanical imaging (Sarvazyan 1998), in which stress patterns are measured on the tissue surface by a pressure sensor array.

MRE can provide measures from deep within the body, at high resolution and in a wide 3D field. Due to the absorption of light in biological tissue, optical based methods are limited to superficial tissue. Likewise, ultrasound methods provide measures at restricted depths due to the limited penetration of ultrasound in tissues, and only for a single direction and over a narrow field. Furthermore, MRE provides access to sites in the body that ultrasound cannot, for example, the brain, for which the skull acts as an acoustic shield. Moreover, MRE can be applied for obese patients or when other factors may prevent ultrasound usage, such as ascites (fluid accumulation in the abdomen). Of course, MRE is not feasible in patients with contraindications to MRI, such as those with pacemakers and cochlear implants, and might not be tolerated by some patients, for example, because of claustrophobia. Ultrasound elastography methods are not affected by these limitations, and furthermore, are faster to implement, less expensive, and more widely available than MRE, but on the down side ultrasound methods are more hampered by imaging noise.

3.4 Background of Magnetic Resonance Imaging

This section will provide a brief description of some of the basic principles behind MRI. For a full explanation, readers are referred to MRI textbooks such as by McRobbie et al. (2007) and Brown et al. (2014).

Magnetic resonance imaging is a diagnostic method based on the phenomenon of nuclear magnetic resonance. In the theory of quantum mechanics, elementary particles, composite particles (hadrons), and atomic nuclei carry a property of intrinsic angular momentum known as spin. Each kind of elementary particle has a particular magnitude of spin, which is indicated by an assigned spin quantum number, I, which can take only half integer or integer values. In MRI, hydrogen (^1H) is the element most commonly focused on, as it is a highly abundant element in biological tissues (e.g., water and fat), although some work is carried out with other elements, including sodium (^{23}Na) and phosphorus (^{31}P). The nucleus of the ^1H atom consists of one proton, which is a composite particle with $I = 1/2$. The intrinsic spin of the proton implies that it possesses angular momentum, $\mathbf{p}$, which is related to I through

$$\mathbf{p} = \hbar \mathbf{I} \tag{3.1}$$

where:
$\hbar = h/2\pi$
h is the Planck constant
$\mathbf{p}$ and $\mathbf{I}$ are vector quantities

The proton also has a unit positive electric charge, and hence, in combination with its angular momentum, it is a *rotating charge*, with an associated magnetic dipole moment, $\mathbf{\mu}$:

$$\mathbf{\mu} = \gamma \mathbf{p} \tag{3.2}$$

where γ is the gyromagnetic ratio, which is a constant for a particular kind of nucleus (e.g., for the ^1H nucleus $\gamma/2\pi = 4258$ Hz/G). Because of the magnetic dipole moment, in the presence of a magnetic field the nucleus behaves similar to a bar magnet, with north and south poles, and will interact with the magnetic field. However, in the quantum mechanical model, $\mathbf{\mu}$ can have only $2I + 1$ orientations with respect to the magnetic field, or $2I + 1$ energy levels. In the instance of the proton with $I = 1/2$, the magnetic dipole moment can align with or against the magnetic field, which are the low- and high-energy states, respectively (Figure 3.3). The energy difference between these two states, ΔE, is proportional to the magnetic field strength, B_0:

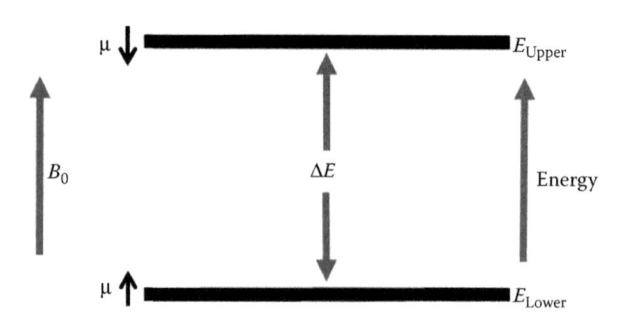

FIGURE 3.3 Energy level of diagram for the magnetic dipole moment of a proton.

$$\Delta E = \frac{\mu B_0}{I} = \gamma \hbar B_0 \tag{3.3}$$

A transition can be induced between the energy states through applying electromagnetic radiation energy (radiowaves) of the appropriate frequency, ω_L, called the Larmor frequency:

$$\omega_L = \frac{\Delta E}{\hbar} = \gamma B_0 \tag{3.4}$$

Hence, a transition can only be made with radiation of the correct resonant frequency, which is proportional to the magnetic field strength. Following excitation to a higher energy spin state, the protons return to the lower energy state through a process called relaxation, with the emission of electromagnetic radiation at the resonant frequency. By altering the local magnetic field strength through the use of magnetic field gradients, the local resonant frequency can be altered. Using this mechanism, the spatial location of the protons can be mapped through observing the frequency of the emitted radiation, and this is the underlying principle of MRI. The relative phase of neighboring proton spins can likewise be altered through application of a magnetic field gradient. The spins of protons that lie in different positions along the magnetic field gradient will rotate at different frequencies, and hence over time become out of phase with respect to each other. The electromagnetic radiation emitted by these protons will contain this phase information; and hence, phase can be used in combination with frequency to locate the position of the emitting protons in two dimensions.

In the MRI scanner, a magnetic field (B_0) of typically 1.5 or 3 Tesla (T) field strength is applied, and various combinations of radiofrequency (RF) excitation pulses are delivered to the patient from radiotransmitter coils in the presence of magnetic field gradients, whereas the emitted radiation is detected by radioreceiver coils. The combinations of excitation pulses and gradients are commonly referred to as pulse sequences. A single implementation of the pulse sequence acquires one portion of the MRI data, and the cycle is repeated at a fixed time interval (repetition time, denoted TR) until all the data are acquired.

MRI can be carried out for a single two-dimensional (2D) image slice with a defined slice width (typically a few millimeters), and also for a 3D volume. The 3D volume data can be obtained by sequentially acquiring multiple 2D slices in a stack, or alternatively by exciting and acquiring data for the full 3D volume. Orthogonal field gradients are applied in three directions to map proton signals in 3D. For a slice-by-slice acquisition, each slice is acquired by applying a magnetic field gradient in the direction of slice stacking and by setting the acquisition hardware to the appropriate frequency range for the slice being acquired. The gradients used to encode the positions of the nuclei are termed *frequency encoding, phase-encoding*, and *slice-selective*, according to the particular direction in space they are applied and the spatial encoding that is used.

The acquired RF signals are analyzed for frequency and phase, and the resulting data are stored as a spatial Fourier Transform, referred to as *k*-space, which at the end of data acquisition is converted into MR images through the application of an inverse

Fourier Transform. The MR images are signal maps, for which the signal strength at a given location is a function of the local concentration of protons (proton density) and other parameters associated with the chemical environment of the protons (the longitudinal relaxation time, T_1) and the interaction of neighboring spins (the transverse relaxation time, T_2). The signal contrast between different tissue types in MR images is caused by variations in these properties, and the pulse sequences are varied to emphasize different contrasts.

3.5 Magnetic Resonance Elastography Methodology

3.5.1 Dynamic Magnetic Resonance Elastography

3.5.1.1 Dynamic Magnetic Resonance Elastography Hardware

With the addition of appropriate technology to deliver the mechanical waves, MRE can be performed on standard MRI scanners, and commercialized MRE technology is currently available from scanner manufacturers. A variety of actuator devices have been explored (Tse et al. 2009), including pneumatic (Yin et al. 2007b), electromechanical (Muthupillai et al. 1995, Braun et al. 2003), and piezoelectric devices (Chen et al. 2005) (Figure 3.4). The amplitude of dynamic MRE vibrations is typically on the order of microns and should be kept low in accordance with recommended vibration safety limits (Ehman et al. 2008).

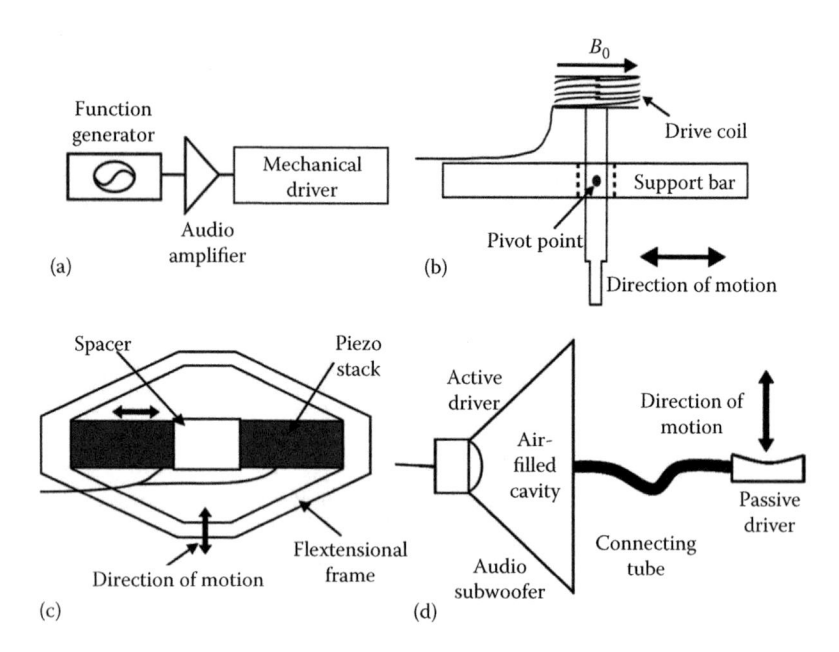

FIGURE 3.4 External driver systems: (a) block diagram of the external driver setup. Examples of typical mechanical drivers include (b) electromechanical, (c) piezoelectric-stack, and (d) pressure-activated driver systems. (Reprinted from Mariappan, Y. K. et al., *Clin. Anat.*, 23, 497–511, 2010. With permission.)

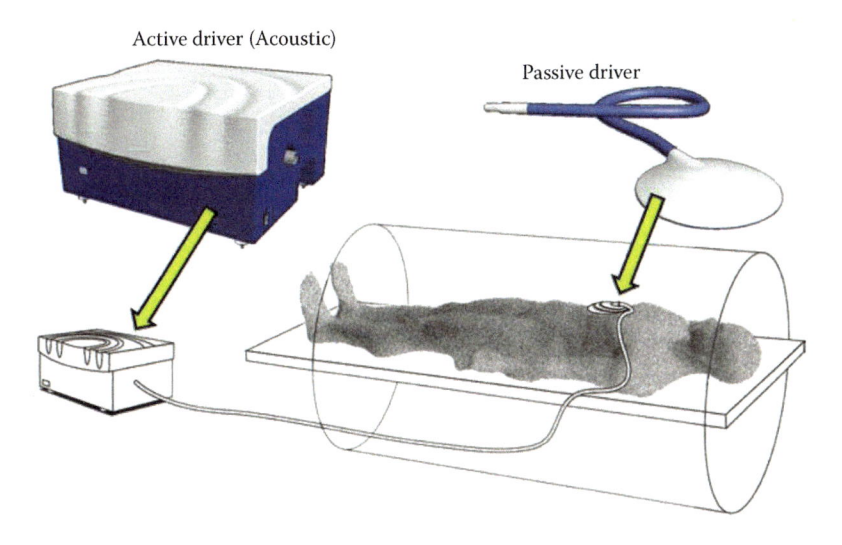

FIGURE 3.5 Illustration of a pneumatic driver system for clinical hepatic MRE. The source of mechanical waves is an *active driver* device that can be located outside the scanner room. Pressure waves are transmitted to a nonmetallic *passive driver* placed in contact with the body by means of a flexible air-filled plastic tube. A flexible membrane on the surface of the passive driver conducts the vibrations into the body to generate propagating shear waves. (Reprinted from Venkatesh, S. K. et al., *J. Magn. Reson. Imaging.*, 37, 544–555, 2013. With permission.)

For pneumatic actuation, a non-MRI compatible driver is positioned outside the scan room, which is commonly a signal generator, connected to an audio amplifier and a loud speaker. This can be connected through an acoustic wave guide (air-filled tube) to an air-filled passive driver placed next to the skin (Figure 3.5), which can take various shapes depending on the application (e.g., drum, disk, or pillow). Alternatively, a rigid rod or piston can be attached to the driver system that delivers vibrations through a rigid passive driver (Asbach et al. 2008, Sack et al. 2009) (Figure 3.6). Longitudinal compression delivered at the skin is mode converted to shear waves at internal tissue boundaries. Pneumatic actuators can suffer from phase delays, which affect synchronization with the MRI sequence, particularly at higher frequencies. However, they have the advantage of allowing high-power amplification in the remote-driver system and are easily adaptable for wave delivery to the site of interest.

Electromechanical actuators consist of electromechanical voice coils, which when positioned in the main magnetic field of the MRI scanner, produce vibrations through the Lorentz force (Braun et al. 2003). Electromechanical actuators can achieve synchronization with the MRI sequence, and produce high-amplitude waves. However, they also tend to cause electromagnetic interference, resulting in eddy currents in the RF receiver coils of the MRI scanner, which cause MRI artifacts and heating effects. As a result, electromechanical drivers must be positioned at a distance from the site of interest. They also have to be placed in a certain orientation with respect to the magnetic field of the MRI scanner.

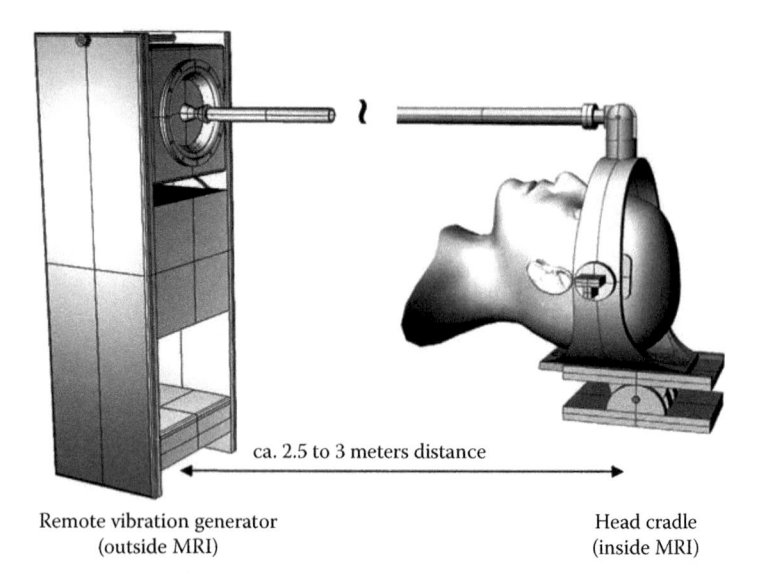

ca. 2.5 to 3 meters distance

Remote vibration generator   Head cradle
(outside MRI)   (inside MRI)

FIGURE 3.6   Head actuator used for stimulating low-frequency shear vibrations in the brain. The audio-driver system is positioned outside the scan room, and is connected through a rigid piston to the head-cradle, which is fitted to the patient lying inside the MRI scanner. (Reprinted from Sack, I. et al., *Neuroimage*, 46, 652–657, 2009. With permission.)

In contrast, piezoelectric actuators (Chen et al. 2005) produce stable mechanical excitation with exact synchronization up to very high frequencies (up to 500 Hz). On the down side, they are more costly than the other types of devices, have relatively low power, and tend to be fragile. Focused ultrasound radiation from ultrasound transducers has also been used for MRE wave delivery (Wu et al. 2000).

3.5.1.2 Dynamic Magnetic Resonance Elastography Pulse Sequences

The MRI pulse sequences employed for dynamic MRE are standard sequences (e.g., spin-echo, gradient-echo, echo-planar imaging, and balanced steady-state free-precession) that are modified to include motion-sensitizing gradients (MSGs), which allow measurement of motion through phase contrast (Figure 3.7). The MSGs are sensitive to the tissue motion in a particular direction and can be applied to capture motion in three orthogonal directions. The MSGs are applied in addition to the frequency and phase-encoding and slice-selective gradients that are used to map the positions of the emitting nuclei in the body.

The MRI scanner is triggered to coincide with the mechanical wave pulses. The spins of the hydrogen nuclei in the tissues are displaced in a cyclic motion by the mechanical waves, and when imaged in combination with the synchronized oscillating MSGs, the motion of the tissue is encoded as phase shifts in the MRI RF signal readout. At a given

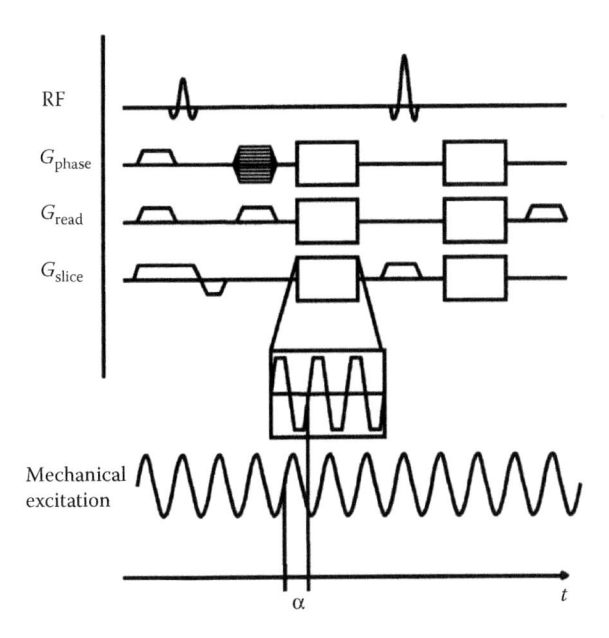

FIGURE 3.7 MRE pulse-sequence diagram. RF = radiofrequency, G_{phase} = phase-encoding gradient, G_{read} = readout gradient or frequency encoding gradient, G_{slice} = slice-selective gradient. Motion sensitizing gradients (MSGs) are inserted in a spin-echo sequence before and after the second RF impulse. α = phase offset between MSG and mechanical excitation. The MSG direction, frequency, and number of cycles are adjustable. (Reprinted from Hamhaber, U. et al., *Magn. Reson. Med.*, 49, 71–77, 2003. With permission.)

position vector (**r**) and relative phase of the MSG and mechanical waves (α), the phase contribution (φ) in the MR image due to the motion and MSG is as follows:

$$\varphi(\mathbf{r},\alpha)=\frac{\gamma NT(\mathbf{G_0}\cdot\mathbf{u_0})}{2}\cos(\mathbf{k}\cdot\mathbf{r}+\alpha) \qquad (3.5)$$

where:
　γ is the gyromagnetic ratio
　N is the number of MSG gradient cycles
　T is the period of the MSG gradient waveform
　$\mathbf{G_0}$ is the MSG vector
　$\mathbf{u_0}$ is the displacement-amplitude vector
　$\mathbf{k}$ is the wave vector

Hence, the phase shift is proportional to the dot product of the gradient and displacement vectors. The phase shifts are unwrapped from the MRI signal and the displacement field is calculated. Usually, several acquisitions (approximately 4–8) are made with different phase offsets α to observe the wave field at different snapshots in time.

3.5.1.3 Dynamic Magnetic Resonance Elastography Inversion Methods

The inversion step calculates the mechanical properties from the measured displacement field and produces a map of the properties, for example, elasticity. Maps of the tissue elasticity are commonly termed *elastograms*. Numerous inversion methods have been developed (Doyley 2012), which employ various assumptions about factors such as the boundary conditions and tissue material properties. These methods have different strengths and weaknesses associated with inherent assumptions, or other factors such as the level of influence of imaging-related noise, which may be amplified through calculations. Generally, inversion methods idealize biological tissues as a continuum and assume a particular material-constitutive model, and many make the assumption, at least within regions, that the tissue is homogeneous, isotropic, and linear viscoelastic. Hence, the greatest challenges for inversion methods occur in heterogeneous, anisotropic, and complex materials, when high-spatial resolution is required, and the MRE data are affected by imaging-related noise.

3.5.1.3.1 Direct Inversion

Direct inversion is one of the more commonly used inversion methods for dynamic MRE. In a harmonic MRE, typically several acquisitions are made to measure the displacement field of the propagating mechanical wave at different snapshots in time. The measured data are processed by Fourier Transform to obtain the frequency domain complex displacement field, $\mathbf{u}$, describing the steady-state as follows:

$$\mathbf{u}(\mathbf{x},t)=\mathbf{u}(\mathbf{x})\exp(i\omega t) \tag{3.6}$$

where:
　　ω is the angular frequency of the mechanical oscillation
　　$\mathbf{x}$ and t are spatial and temporal coordinates

If the tissue is modeled with a locally homogeneous isotropic viscoelastic material-constitutive model, these displacements $\mathbf{u}$ are employed to solve for the viscoelastic parameters using the Navier–Stokes equation for the propagation of an acoustic wave in a viscoelastic solid (Sinkus et al. 2005):

$$\rho\frac{\partial^2 \mathbf{u}}{\partial^2 t}=\mu\nabla^2\mathbf{u}+(\lambda+\mu)\nabla(\nabla.\mathbf{u})+\zeta\frac{\partial\nabla^2\mathbf{u}}{\partial t}+(\xi+\zeta)\frac{\partial\nabla(\nabla.\mathbf{u})}{\partial t} \tag{3.7}$$

where:
　　ρ is the material density
　　λ is the first Lamé parameter describing the material elasticity with regard to the
　　　　compressional wave component
　　μ is the second Lamé parameter describing the material elasticity with respect to the
　　　　shear-wave component (*Note*: This parameter is distinct from the magnetic
　　　　dipole moment, $\boldsymbol{\mu}$)
　　ξ is the viscosity of the compressional component
　　ζ is the shear viscosity

In a near-incompressible medium such as biological tissue, the wavelength of the compressional component is typically very long (meters), whereas the amplitude is very small; and hence, this component is very difficult to measure accurately. Furthermore, the compressional wave velocity tends to differ little between biological tissue types (typically 1540 m/s), whereas the shear-wave velocity can differ substantially (1–10 m/s). Therefore, in dynamic MRE, the shear-wave component of the displacement field is typically used for measurement and characterization, where the task is to estimate μ and ζ. In near-incompressible material, the gradient of the displacement field $\nabla \mathbf{u}$ is very close to zero, perhaps indicating that one could neglect the second and fourth terms on the right-hand side of Equation 3.7. However, though the very small compressional viscosity ξ might make it acceptable to ignore the fourth term, it is not safe to neglect the second term, as the large λ of the near-incompressible material balances the small $\nabla(\nabla \cdot \mathbf{u})$. A preferred approach is to remove the contributions of the compressional wave component from Equation 3.7, and one way to achieve this is by calculating the curl (divergence-free part) of the vector field $(\nabla \times \mathbf{u})$ (Sinkus et al. 2005).

Replacing $\mathbf{u}$ with $\mathbf{v} = \nabla \times \mathbf{u}$, Equation 3.7 becomes

$$\rho \frac{\partial^2 \mathbf{v}}{\partial^2 t} = \mu \nabla^2 \mathbf{v} + \zeta \frac{\partial \nabla^2 \mathbf{v}}{\partial t} \tag{3.8}$$

Replacing $\mathbf{u}$ with $\mathbf{v}$, Equation 3.6 becomes

$$\mathbf{v}(\mathbf{x},t) = \mathbf{v}(\mathbf{x})\exp(i\omega t) \tag{3.9}$$

If this expression for $\mathbf{v}$ is substituted into Equation 3.8 and the derivatives are calculated, a Helmholtz equation is obtained for the steady-state as follows:

$$-\rho\omega^2 \mathbf{v} = \mu \nabla^2 \mathbf{v} + i\omega\zeta \nabla^2 \mathbf{v} \tag{3.10}$$

For 3D MRE data, by separating the components for the three dimensions, one obtains three equations for two unknowns, μ and ζ. By applying the assumption of local homogeneity of the material properties over the neighborhood of the voxel for derivative calculation, the equations can be solved separately for each voxel, using, for example, a least squares calculation (Sinkus et al. 2005). If the tissue is presumed to have a near-zero or negligible viscosity ($\zeta \sim 0$), Equation 3.10 is simplified to

$$-\rho\omega^2 \mathbf{v} = \mu \nabla^2 \mathbf{v} \tag{3.11}$$

This approach to inversion has also been described as algebraic inversion of the differential equation (AIDE) (Manduca et al. 2001). However, a full AIDE inversion described in Oliphant et al. (2001) solves for both Lamé coefficients (λ and μ) for the compressional and shear-wave components.

As calculation of the curl can result in amplification of imaging-related noise, and also requires measurement of three displacement components, which may entail prohibitively long patient scan times, some investigators have used high-pass filters to suppress the low-frequency compression wave components, while maintaining the high-frequency shear-wave components. This approach can be applied for 2D, and even 1D MRE data; however, only an approximate separation of wave components is achieved, and this may cause errors in the inversion. In addition for 2D data, when using a high-pass filter, an assumption is made that the shear waves are propagating parallel to the plane of the imaging slice, and if the direction of the wave is oblique to the slice, this may lead to errors in the inversion estimates.

The variational inversion method described in Romano et al. (1998) avoids derivative calculations in solving the Lamé coefficients (μ and λ) by using the weak or variational form of the Navier–Stokes equation with appropriately chosen smooth test functions, and this approach allowed local inhomogeneity of the material parameters. Honarvar et al. (2013) developed a curl-based finite-element reconstruction direct-inversion method to solve for μ, which again used a weak formulation of the equation of motion, thereby allowing for local inhomogeneity of μ.

The shear-wave parameters μ and ζ are related to the complex shear-modulus parameter $G^*(\omega)$, which relates complex stress $\sigma^*(\omega)$ to complex strain $\epsilon^*(\omega)$

$$\sigma^*(\omega) = G^*(\omega)\epsilon^*(\omega) = \left[G'(\omega) + iG''(\omega) \right] \epsilon^*(\omega) \tag{3.12}$$

where:
G' is the storage modulus
G'' is the loss modulus
$G'(\omega) = \mu$ and $G''(\omega) = \omega\zeta$

Hence, some MRE studies quote the viscoelastic measures in terms of G' and G'' (Sack et al. 2009). Viscoelasticity can also be defined by the magnitude of the complex shear modulus ($|G^*|$) in combination with the phase lag between stress and strain δ (which is $0°$ for purely elastic material and $90°$ for purely viscous material), where $G' = |G^*|\cos\delta$ and $G'' = |G^*|\sin\delta$.

Most applications of direct inversion assume isotropic material properties; however, Sinkus et al. (2000) developed a tensor-based direct-inversion approach to detect breast tumors based on the anisotropy of Young's Modulus. For a transient MRE technique, McCracken et al. (2005) also employed a direct-inversion method and compared it with a time of arrival method that measured wave speed to estimate shear stiffness.

3.5.1.3.2 *Local-Frequency Estimation*

Another commonly used inversion approach for dynamic MRE is the local-frequency estimation (LFE). In this method, the local spatial frequency, f_{sp}, of the shear-wave propagating pattern is measured. This can be estimated using an algorithm that combines measures over multiple scales obtained from wavelet filters (Manduca et al. 2001). The local spatial frequency is related to the local shear-wave speed c_s, and the mechanical driving frequency f_{mech}, that is, $c_s = f_{mech}/f_{sp}$. c_s is also related to the complex shear modulus G^*:

$$c_s = \sqrt{\frac{2|G^*|^2}{\rho\left(\text{Re}(G^*) + |G^*| \right)}} \tag{3.13}$$

where $Re(G^*)$ is the real component of G^* (i.e., the storage modulus G'). For a material model with zero viscosity, this equation becomes

$$\rho c_s^2 = G \qquad (3.14)$$

where G is the shear modulus for which the imaginary component (the loss modulus, G'') is zero. The approximation of zero viscosity can be made for some tissues with low dispersion and attenuation. In this instance, G can be estimated from c_s^2 as an effective *shear stiffness* (i.e., with the assumption of ρ is equal to that of water, 1000 kg/m³). LFE is relatively robust to imaging-related noise and can be implemented in 3D. However, it assumes local homogeneity, isotropy and incompressibility, has limited spatial resolution and is prone to blurring at tissue boundaries. The phase-gradient method (Catheline 1999) is similar to the LFE method; however, this time the local frequency is estimated from the gradient of the phase of the measured harmonic oscillation.

3.5.1.3.3 Other Dynamic Magnetic Resonance Elastography Inversion Methods

The inversion problem can be treated as an iterative parameter optimization problem by simulating the MRE displacement field with a set of material property parameters and comparing this with the acquired MRE data. For example, Van Houten et al. (1999) (2003) have developed an iterative finite element model (FEM)-based methodology to simulate the displacement fields, and to make the problem tractable for large datasets, the data were separated into overlapping subzones. Using this inversion method, parameters have also been calculated for a poroelastic material-constitutive model, that is, a solid elastic matrix permeated by fluid (Perrinez et al. 2010).

Other investigators have sought to characterize viscoelastic biological tissue according to its variable response to different mechanical frequencies, and have measured the complex shear modulus or wave speed at different frequencies. Multifrequency measurements have been fit to candidate rheological models, such as the Voigt and Spring–Pot models (Klatt et al. 2007, Sack 2009), and used to characterize the exponent of a power law describing the behavior of the complex shear modulus (Sinkus et al. 2007). Multifrequency measurements have also been summarized in combined multifrequency parameters, such as multifrequency dual elastovisco inversion (MDEV) (Reiss-Zimmermann et al. 2014) and multifrequency wave-number recovery (Tzschatzsch et al. 2016). Postprocessing algorithms for reducing the effects of imaging-related noise have also been explored (Barnhill et al. 2016).

3.5.2 Quasi-Static Magnetic Resonance Elastography

3.5.2.1 Quasi-Static Magnetic Resonance Elastography Hardware

For an *in vivo* breast study, a quasi-static MRE device was constructed that attached to the breast RF coil of the MRI scanner, and breast tissue was compressed between two parallel plates, with one plate moving in and out in a sinusoidal motion as driven by an ultrasonic motor at a frequency of 1 Hz (Plewes et al. 2000).

For an *in vitro* application (McGrath et al. 2012), a quasi-static MRE system was built for *ex vivo* prostate whole specimens to determine the effect of exposure to pathology

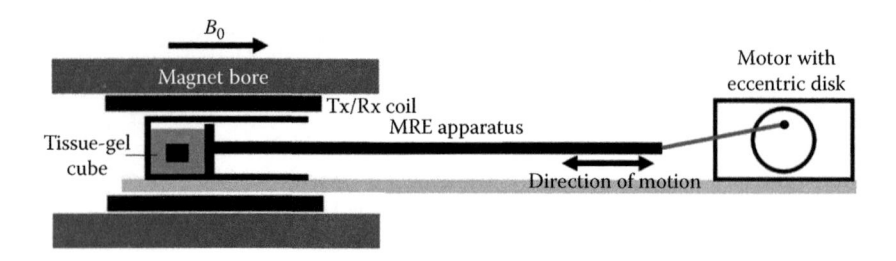

FIGURE 3.8 Schematic of the quasi-static MRE device positioned within the scanner bore consisting of a sample holder (length = 20 cm, inside cross-sectional area = 8 × 8 cm²) in which the samples are compressed by a square plate (8 × 8 cm² surface area), which is attached through a mechanical piston to the motor. (Reprinted from McGrath, D.M. et al., *Magn Reson Med.*, 68, 152–165, 2012. With permission.)

fixative preserving solution (formaldehyde) on tissue elasticity. An apparatus was built in acrylic consisting of a holder and a compression plate, which was attached to a piston connected to an ultrasonic motor through an eccentric disk (Figure 3.8). The device was positioned in the bore of a 7-T preclinical MRI scanner, which was triggered by the motion of the piston. The tissue specimens were embedded in a block of gelatin before scanning to provide stability during cyclic compression.

3.5.2.2 Quasi-Static Magnetic Resonance Elastography Pulse Sequences

For quasi-static MRE, the pulse sequences used include the spin-echo method (Plewes et al. 1995) and the stimulated echo-acquisition mode (STEAM) method (Chenevert et al. 1998, Plewes et al. 2000, McGrath et al. 2012) (Figure 3.9). The STEAM pulse

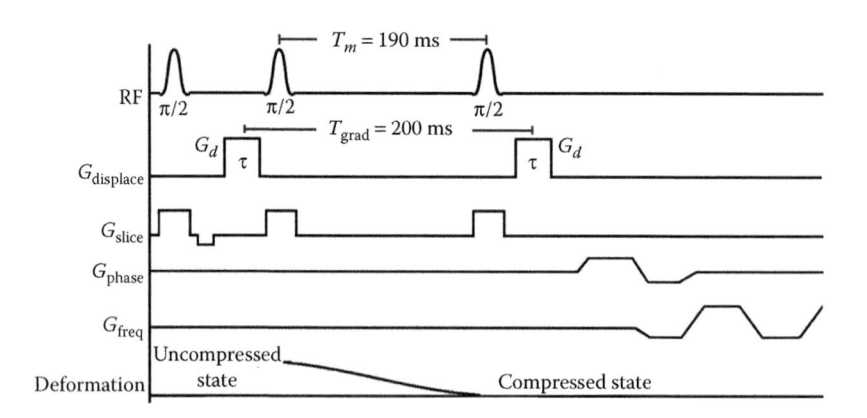

FIGURE 3.9 Pulse-sequence diagram of a STEAM sequence for quasi-static MRE, with displacement-encoding gradients and EPI readout. The mechanical displacement from the uncompressed to compressed states occurs during the mixing time, T_m. $G_{displace}$ = displacement-encoding gradient. (Reprinted from McGrath, D.M. et al., *Magn Reson Med.*, 68, 152–165, 2012. With permission.)

sequence includes three RF pulses, and for MRE two displacement-encoding gradients are applied: (1) one before the second RF pulse, and (2) another after the third RF pulse. The displacement-encoding gradients are specified by the following parameters: gradient amplitude (G_d) and gradient duration (τ), which determine the displacement sensitivity ($\Phi_d = \gamma G_d \tau$) and the gradient pulse-separation interval (T_{grad}). The transition between the uncompressed and compressed states is timed to occur during the mixing time, T_m, between the second and third RF pulses. In McGrath et al. (2012), echo-planar imaging (EPI) was incorporated into the pulse sequence, to facilitate a more rapid acquisition, as the application to pathology specimens was time sensitive, to limit tissue degradation before histopathological analysis. The pulse sequence was repeatedly applied to acquire the MRE data for the full-imaging volume step-by-step, and the time to cycle through the pulse sequence (TR) was set to match the compression cycle time (1 s).

3.5.2.3 Quasi-Static Magnetic Resonance Elastography Inversion Methods

The displacement and strain patterns measured by elastography are related to the mechanical properties of the tissues, and for some applications visualization of the measured strain field in 2D or 3D may be sufficient to assess variations in elasticity, such as in the study by Hardy et al. (2005) with *ex vivo* cartilage. However, the strain field is also related to the deformational geometry, and when more precise measurement of the elastic modulus is required, a full-inversion approach is adopted.

3.5.2.3.1 *Direct Inversion*

In the static or quasi-static case, the time-derivative terms in the Navier–Stokes equation are zero. For an incompressible material, the term $\lambda \nabla \cdot u$ is indeterminate. This term can be replaced by a parameter p, defined as average stress or pressure (Bishop et al. 2000, Plewes et al. 2000):

$$\nabla p + \mu\left(\nabla^2 u + \nabla\nabla \cdot u\right) + \left(\nabla\mu\right)\left(\nabla u + \nabla u^{\mathrm{T}}\right) = 0 \qquad (3.15)$$

Bishop et al. (2000) and Plewes et al. (2000) employed a direct-inversion method to solve this equation for μ and p, by discretizing Equation 3.15 and solving with linear equation algorithms, that is, singular value decomposition with Tikhonov regularization. This method requires defined boundary conditions for the unknown elastic modulus, and in Bishop et al. (2000), Plewes et al. (2000), a constant boundary modulus was assumed for a relative modulus reconstruction, that is, modulus values were constructed as ratios of the constant boundary value. This method assumes linear elasticity, small strain deformation, and 2D plane strain conditions, and these approximations make the matrix equation computationally tractable, with fewer unknown variables and spatial locations. Chenevert et al. (1998) also carried out a direct inversion for quasi-static MRE, but applied another partial derivative to remove the pressure term. However, by including the pressure term as an unknown (Bishop et al. 2000, Plewes et al. 2000) avoided applying third-order derivatives, which can lead to amplification of imaging-related noise.

3.5.2.3.2 Iterative Finite Element Model-Based Inversion

Tissue deformation in static or quasi-static elastography can be deemed governable by equations of equilibrium and strain displacement. The equilibrium equation of a continuum under static loading is therefore applicable as follows:

$$\frac{\partial \sigma_{i1}}{\partial x_1} + \frac{\partial \sigma_{i2}}{\partial x_2} + \frac{\partial \sigma_{i3}}{\partial x_3} + f_i = 0 \quad i = 1,2,3 \tag{3.16}$$

where:

σs is the stress-tensor components
f_i is the body forces per unit volume
$X = (x_1, x_2, x_3)$ are Cartesian coordinates

For a linear elastic material under static deformation, the strain (ε_{ij}) tensor is as follows:

$$\varepsilon_{ij} = \frac{1}{2}\left(\frac{\partial u_i}{\partial x_j} + \frac{\partial u_j}{\partial x_i}\right) \tag{3.17}$$

where u_i is a component of the displacement vector $U = (u_1, u_2, u_3)$. If compression is limited so that the strains do not exceed 0.05, the deformations can generally be assumed linearly elastic, and Equation 3.17 applies. For an isotropic linear elastic material, the mechanical strains and stresses are related according to Hooke's law:

$$\sigma_{ij} = \frac{E}{1+v}\left(\varepsilon_{ij} + \frac{v}{(1-2v)}\delta_{ij}\varepsilon_{kk}\right) \tag{3.18}$$

where:

δ_{ij} is the Kronecker delta symbol
E and v are the Young's modulus, and the Poisson's ratio of the material, respectively

Plewes et al. (2000) and Samani et al. (2001) presented an iterative inversion method for quasi-static MRE, which involved solving the so-called *forward problem*, by generating a simulated stress distribution from an estimated distribution of Young's modulus. The forward solution was generated using a FEM-based simulation, which was a 3D contact problem with displacement-boundary conditions specified by the plates of the compression device. At each iteration, E was updated according to an equation derived from Hooke's law:

$$\frac{1}{E^{i+1}} = \frac{\varepsilon_{11}^{MRE}}{\sigma_{11}^i - v\sigma_{22}^i - v\sigma_{33}^i} \tag{3.19}$$

where the updated modulus E^{i+1} of each finite element is calculated from one component of the MRE-measured strain (ε_{11}^{MRE}, the component parallel to the direction of motion of the compression device) and the simulated three normal stress components of the previous iteration, combined with the assumed value for v (for the assumption of near-incompressibility v was estimated at 0.499). The starting guess for E was estimated from the reciprocal of the measured strain. For the convergence metric, the mean squared proportional differences between consecutive modulus updates were compared with a defined tolerance (tol):

$$\left\| \frac{E^{i+1} - E^i}{E^i} \right\| < \text{tol} \tag{3.20}$$

and tol was set to an arbitrary small number. To stabilize the iterative calculations in the presence of imaging-related noise, Plewes et al. (2000), Samani et al. (2001) applied a geometric constraint that used a priori knowledge from other MR imaging to segment the tissue into different types, for example, for the breast: fat, fibroglandular tissue, and tumor. At each iteration, the new estimates for E were averaged across each tissue region. The final reconstruction gave the relative average modulus values of the different inner regions with respect to the modulus of the outer region. Without knowledge of the true magnitude of the applied forces, the simulation produced a stress distribution with an arbitrary scaling, and hence the overall scaling of the estimated E values was also arbitrary. However, the relative distribution of modulus values will still be valid for the specified geometry and boundary conditions. As the algorithm only required a single component of strain (the normal component, ε_{11}), only a single-motion encoding direction was needed in the direction of motion. Assuming a linear elastic material, ε_{11} was calculable as $\partial u_1 / \partial x_1$, which was calculated from the local MRI signal phase and the displacement sensitivity, Φ_d.

In McGrath et al. (2012) (Figure 3.10), this methodology was adapted to measure E in *ex vivo* tissue specimen volumes before and after exposure to pathology fixative solution, and it was sought to determine the variable effects with distance from the specimen surface and with fixation time. The tissue was embedded in a block of gel before imaging, and the gel elastic modulus was characterized separately by mechanical indentation testing. As the gel material properties were expected to be uniform, the gel E estimates were averaged at each iteration, whereas the tissue E estimates were allowed to vary freely for each finite element. Simulations were carried out to verify the accuracy of this approach in the presence of different levels of imaging noise. At the final iteration, a ratio map was produced of the tissue element E values as a ratio of the calculated average gel E value. These ratio-E values were converted to absolute values through multiplication with the indentation-measured E value for the gel.

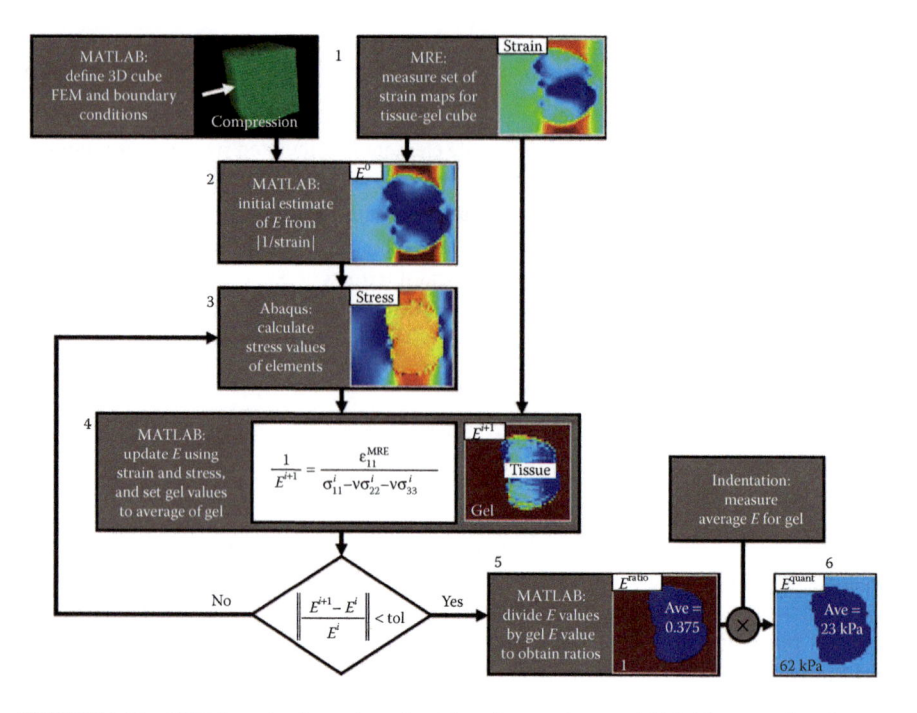

FIGURE 3.10 FEM iterative inversion algorithm for quasi-static MRE (showing data from a simulation experiment): (1) finite elements and boundary conditions defined, and MRE used to measure strain; (2) strain used to estimate the initial E values; (3) ABAQUS (finite element solver) software simulates the compression and calculates three components of stress per element; (4) MRE-measured strain combined with calculated stress to provide new estimates of E and the gel E values are reset to their average, and convergence tested; (5) once converged, values are divided by the average gel value to obtain ratio-E values; and (6) multiplication by the indentation-measured gel E value to obtain quantitative E measures for tissue elements. (Reprinted from McGrath, D.M. et al., *Magn Reson Med.*, 68, 152–165, 2012. With permission.)

3.6 Applications of Magnetic Resonance Elastography

MRE has been applied to a wide range of organs and tissue types in humans and in animal models, both *in vivo* and *in vitro*. This section will provide an overview of MRE applications to date.

3.6.1 Clinical Applications

In vivo human MRE studies have been carried out for multiple organs and tissues, including liver (Huwart et al. 2006, 2007, 2008, Rouviere et al. 2006, Klatt et al. 2007, Godfrey et al. 2012, Venkatesh et al. 2013, Su et al. 2014, Singh et al. 2015); spleen (Talwalkar et al. 2009); kidneys (Rouviere et al. 2011, Lee et al. 2012, Low et al. 2015b); uterus and cervix (Jiang et al. 2014); pancreas (Shi et al. 2015, Venkatesh and Ehman 2015, An et al. 2016,

Dittmann et al. 2016, Itoh et al. 2016); breast (Plewes et al. 2000, Sinkus et al. 2000, 2005, McKnight et al. 2002, Van Houten et al. 2003, Sinkus et al. 2007, Hawley et al. 2016); brain (Klatt et al. 2007, Sack et al. 2008, 2009, Di Ieva et al. 2010, Murphy et al. 2011, 2016, Streitberger et al. 2011, 2012, 2014, Guo et al. 2013, Lipp et al. 2013, Braun et al. 2014, Romano et al. 2014); muscle (Basford et al. 2002, Sack et al. 2002, Ringleb et al. 2007); heart (Elgeti et al. 2009, Robert et al. 2009, Sack et al. 2009, Arani et al. 2016); prostate (Kemper et al. 2004, Li et al. 2011, Arani et al. 2013, Sahebjavaher et al. 2013); abdominal aorta (Kolipaka et al. 2012, Xu et al. 2013, Damughatla et al. 2015, Kenyhercz et al. 2016, Kolipaka et al. 2016); lungs (Mariappan et al. 2011, 2014); and head and neck (Bahn et al. 2009, Yeung et al. 2013). The following sections will expand on some of these clinical applications.

3.6.1.1 Liver

Although most clinical MRE applications have been initial explorations of method feasibility and sensitivity to disease, liver MRE is a notable exception, as it has already been widely adopted in clinical practice for the diagnosis of chronic liver disease (CLD), such as fibrosis (Figure 3.11). Although biopsy is the gold standard diagnostic for liver fibrosis, it also carries the risks of an invasive procedure and is sometimes not tolerated by patients. In addition, biopsy can be prone to error through sampling at the wrong locations, or if the sample volume is insufficient or of poor quality. MRE could be used instead of biopsy (Venkatesh et al. 2013) for longitudinal monitoring of disease progression or treatment, and also for guiding the selection of biopsy samples when histopathology is required (Perumpail et al. 2012). A number of studies have determined MRE to be a highly accurate diagnostic tool for CLD, with increases in liver stiffness accompanying progression through stages of fibrosis to cirrhosis (Yin et al. 2007a, Su et al. 2014, Singh et al. 2015). A recent comparison of MRE with diffusion-weighted MRI for

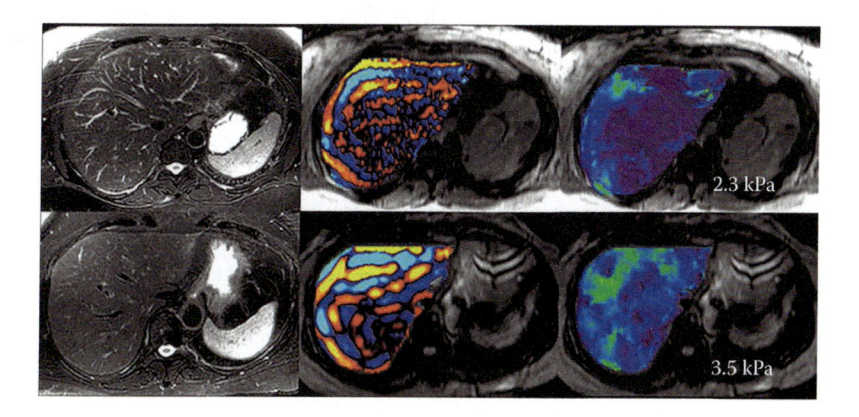

FIGURE 3.11 Early detection of fibrosis with MRE. Two patients with chronic hepatitis B. The patient in the upper row has normal liver stiffness, whereas the patient in the lower row has modestly elevated liver stiffness. Biopsies excluded fibrosis in the first patient and showed mild fibrosis in the second. All conventional MR images were normal in both patients. (Reprinted with permission from Venkatesh, S.K. et al., *J. Magn. Reson. Imaging.*, 37, 544–555, 2013. With permission.)

liver fibrosis, steatosis, and inflammation demonstrated that MRE was more sensitive to the effects of fibrosis than the apparent diffusion coefficient (ADC), whereas MRE was less sensitive to steatosis than ADC (Leitao et al. 2016). In a retrospective study for patients with nonalcoholic fatty liver disease (NAFLD), Chen et al. (2011) reported that NAFLD patients with inflammation in the absence of fibrosis had higher liver stiffness than those with steatosis and lower stiffness than those with fibrosis. MRE has also been demonstrated as effective in distinguishing benign and malignant liver tumors with malignant lesions having a higher mean stiffness (Venkatesh et al. 2008, Hennedige et al. 2016). Furthermore, a recent MRE study determined that liver biomechanics varied between children and adults (Etchell et al. 2016).

3.6.1.2 Spleen and Kidneys

In an MRE study by Talwalkar et al. (2009), it was found that patients with CLD had a higher splenic stiffness compared with controls, and Shin et al. (2014) found a positive linear correlation between splenic stiffness and hepatic stiffness and grade of esophageal varices. It has also been found by Guo et al. (2015) that viscoelastic properties of the spleen as measured with a multifrequency MRE method depended on hepatic venous pressure gradient (HVPG) measurements before and after the placement of a transjugular intrahepatic portosystemic shunt.

The number of renal MRE studies to date has been limited (Rouviere et al. 2011, Low 2015b), but the findings so far suggest a dependency of renal stiffness on liver fibrosis (Lee et al. 2012) and hepatorenal syndrome (HRS) (Low et al. 2015a).

3.6.1.3 Brain

MRE has introduced the possibility of estimating *in vivo* brain tissue mechanics, as the skull acts as a barrier that precludes the use of other methods, such as palpation (except during surgery) and ultrasound-based elastography. Initial investigations with brain MRE have found that cerebral biomechanical properties are dependent on age and gender (Arani et al. 2015, Sack 2009) and are sensitive to a variety of neurodegenerative and neurological diseases, including Alzheimer's disease (AD) (Murphy et al. 2011, 2016) (Figures 3.12 and 3.13), Parkinson's disease (Lipp et al. 2013), hydrocephalus pre- and post-shunt placement (Streitberger et al. 2011, Freimann et al. 2012, Fattahi et al. 2016), and multiple sclerosis (Wuerfel et al. 2010, Streitberger et al. 2012). Some initial MRE work with brain cancer patients has indicated that MRE may be used to distinguish different types of intracranial neoplasms, such as meningiomas and intra-axial neoplasms (Reiss-Zimmermann et al. 2014), and meningiomas were found to be stiffer than pituitary adenomas (Sakai et al. 2016).

However, the cerebral biomechanic measures for healthy volunteers have varied widely between different studies, that is, in a review by Di Ieva et al. (2010) it was found that for brain white matter the shear modulus varied between 2.5 and 15.2 kPa, and for gray matter between 2.8 and 12.9 kPa. Moreover, the expected impact on cerebral biomechanics of neurodegenerative diseases is low; for example, Murphy et al. (2011) reported only a 7% reduction in brain stiffness for Alzheimer's patients compared with healthy controls. Although the different MRE measures for healthy brain might reflect heterogeneity across the population, recent studies have explored other potential influences, such as

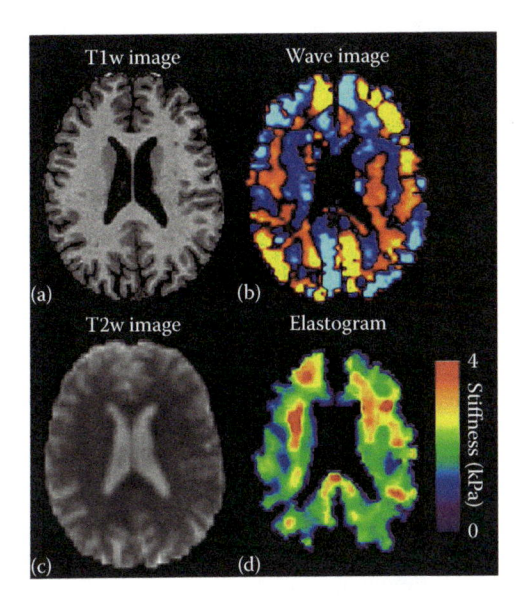

FIGURE 3.12 Example MRE images from a cognitively normal control. The MRI T1-weighted (T1w) image is shown in (a), and the T2-weighted (T2w) MRE magnitude image is shown in the (c). A curl wave image is shown in (b), along with the resulting elastogram in the (d) panel. (Reprinted from Murphy, M.C. et al., *NeuroImage: Clinical*, 10, 283–290, 2016.)

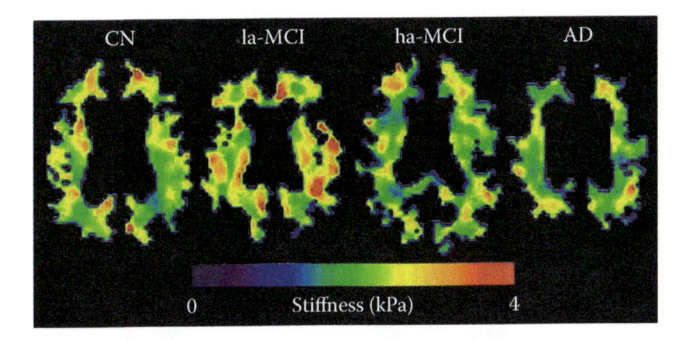

FIGURE 3.13 Example elastograms across the Alzheimer's disease (AD) spectrum. Relative to a cognitively normal (CN) control subject, stiffness is elevated in low (but still positive)-amyloid subjects with MCI (la-MCI) before falling in high-amyloid subjects with MCI (ha-MCI) to levels that are common within the AD group. (MCI = mild cognitive impairment). (Reprinted from Murphy, M.C. et al., *NeuroImage: Clinical.*, 10, 283–290, 2016. With permission.)

interference patterns from wave reflection and scattering at brain tissue boundaries. In McGrath et al. (2015), using a FEM-based brain MRE simulation, it was predicted that reflections from brain tissue interfaces with cranial features such as the falx cerebri could lead to inversion error artifacts on the order of 10%–20%, and that these errors would vary between individuals. Another potential influence on the variability of the healthy baseline data is variation in the methodology between different studies, including the means of mechanical wave delivery to the brain through the skull, for example, bite-bar (Green et al. 2008), head-cradle (Sack et al. 2009), acoustic pillow (Murphy et al. 2011), and through the mechanical vibrations produced in the patient bed from the MRI scanner (Gallichan et al. 2009). Different wave-delivery methodologies were compared in McGrath et al. (2016) using FEM simulations, and simulated displacements fields and inversions varied greatly between wave-delivery methods, with the inversion error differing by as much as 11% between methods. Furthermore, in a recent brain MRE acquisition study (Fehlner et al. 2015), the head-cradle method of Sack et al. (2009) was compared with a newer remote excitation method, and it was found that in the brain regions examined the magnitude and phase of the complex shear modulus differed by up to 6% and 13%, respectively. Another potential cause of variability in the baseline healthy MRE data is the influence of anisotropic material properties, and a recent study found that variation in the direction of head actuation could result in differences in mechanical property estimates (as much as 33%), and particularly in the areas of high anisotropy, as determined by diffusion-tensor MR imaging (Anderson et al. 2016). Hence, priority should be given in future brain MRE development to the examination of the influence of variable-wave delivery, to determine an optimum methodology.

3.6.1.4 Prostate

In vivo MRE methods for prostate are still in development with different wave-delivery methods being explored, that is, through the rectum, perineum, and pubic bone (Kemper et al. 2004, Li et al. 2011, Arani et al. 2013, Sahebjavaher et al. 2013, 2014) (Figure 3.14). So far, prostate MRE studies have demonstrated a potential utility to diagnose and locate prostate cancer, and distinguish cancer from benign prostatitis (Li et al. 2011). The accuracy of prostate cancer diagnosis with MRE has been evaluated through comparison of MRE of *ex vivo* whole prostatectomy specimens with gold-standard histology (McGrath et al. 2011, 2017, Sahebjavaher et al. 2015).

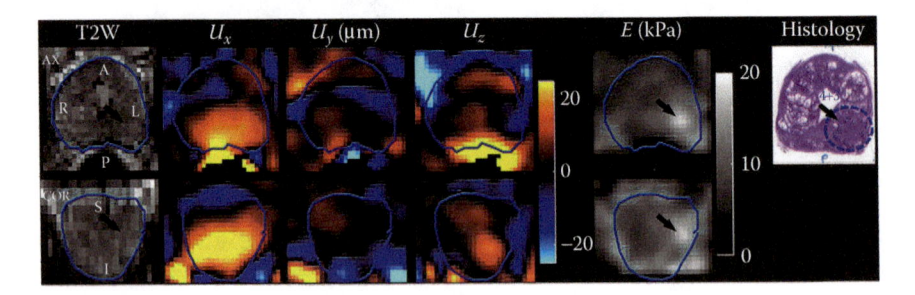

FIGURE 3.14 The images show the results of the transperineal motion-encoding gradient (MRE) approach in a patient prior to a radical-prostatectomy procedure. The waves are present in all three directions in the entire gland, demonstrating the effectiveness of the transducer for inducing the waves, and the pulse sequence for encoding the resulting displacements. In the reconstructed elasticity image, the Gleason 4 + 3 tumor can be identified (see matching histopathology slide). (Ax = axial imaging slice, Cor = coronal imaging slice). (Reprinted from Sahebjavaher, R.S. et al., *NMR Biomed.*, 27, 784–794, 2014. With permission.)

3.6.1.5 Breast

Clinical applications of MRE for the breast are still under development, and the different methodologies explored include the quasi-static MRE method (Plewes et al. 2000), and harmonic MRE with an electromechanical driver (Sinkus et al. 2000 and 2005, McKnight et al. 2002) and a soft actuator connected to an acoustic driver (Hawley et al. 2016). So far, breast MRE has demonstrated promising potential as a means of diagnosing and localizing breast cancer (Sinkus et al. 2005) and for differentiating benign and malignant lesions (Lorenzen et al. 2002, Sinkus et al. 2007). Breast MRE has also been explored in combination with contrast enhanced MRI to improve breast cancer diagnosis (Siegmann et al. 2010).

3.6.1.6 Musculoskeletal

A wide range of musculoskeletal MRE studies have been carried out to measure muscle stiffness, and recent developments include implementation at high magnetic field strength (3 T) to facilitate shorter patient scan times (Hong et al. 2016), and a pneumatic actuator that is adaptable to different shoulder shapes for examining supraspinatus muscle (Ito et al. 2016). The musculoskeletal MRE applications so far have included measurement for muscle cartography (Debernard et al. 2013) (Figure 3.15), muscle development from childhood to adulthood (Debernard et al. 2011), muscle aging (Domire et al. 2009), muscle

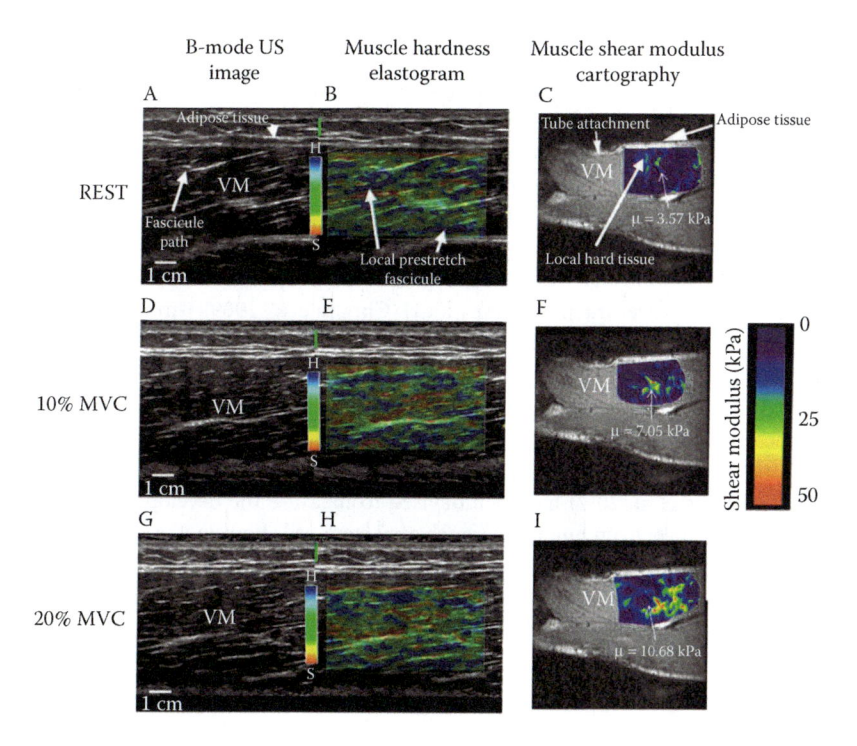

FIGURE 3.15 A–I: MR and ultrasound elastographic images recorded in the same vastus medialis (VM) muscle of an adult in passive and active (10% and 20% of MVC) states. MVC = Maximum voluntary contraction. (Reprinted from Debernard, L. et al., *Muscle Nerve*, 47, 903–918, 2013. With permission.)

activity (Heers et al. 2003), muscle contraction measurement (Jenkyn et al. 2003), neuromuscular dysfunction (Basford et al. 2002), myositis (McCullough et al. 2011), muscle stiffness before and after treatment of hyperthyroidism (Bensamoun et al. 2007), and hormonal therapy (Brauck et al. 2007). Hence, MRE has a wide utility for musculoskeletal medicine and may become an important future diagnostic method.

3.6.2 Preclinical Applications

For many of the MRE clinical studies, prior developmental work was carried out in animal models, and some preclinical MRE methods have yet to be translated to humans. For example, for lungs, an MRE methodology was developed with hyperpolarized helium (^{3}He) gas using *ex vivo* porcine lung (McGee et al. 2008). In another pulmonary application, proton (^{1}H) MRE was carried out in *ex vivo* rat lung to explore measurement of normal and edematous ventilator-injured lung (McGee et al. 2012).

For a cardiac MRE method, a model of myocardial hypertension with *in vivo* porcine heart was employed (Mazumder et al. 2016). MRE for the eye was explored using an *ex vivo* bovine globe model (Litwiller et al. 2010), and applications for cartilage were tested using *ex vivo* bovine cartilage with dynamic MRE (Lopez et al. 2008) and quasi-static MRE (Hardy et al. 2005).

In vivo mouse brain MRE methods were developed to measure demyelination *in vivo* in a murine model of multiple sclerosis (Schregel et al. 2012), and to measure the effects of stroke on brain viscoelastic properties (Freimann et al. 2013); and in both studies the MRE data were compared with histology. A microscopic MRE method was developed for *ex vivo* rat brain to study a traumatic brain injury model (Boulet et al. 2011), and MRE was employed to observe the development of postnatal rat brain in Pong et al. (2016). Furthermore, MRE methodology was developed for *in vivo* ferret brain, incorporating a piezoelectric device that delivered waves through a bite-bar (Feng et al. 2013).

For prostate MRE, various wave-delivery routes were explored using different animal models: through the urethra in a canine model (Chopra et al. 2009); through the rectum in a porcine model (Thormer et al. 2013). For liver MRE, a needle shear wave driver was developed and applied to a murine model of liver fibrosis (Yin et al. 2007c), and such a preclinical methodology could be used in longitudinal studies to understand disease progression or assess treatment effects.

For a tissue engineering application, a microscopic MRE methodology (Othman et al. 2005, Curtis et al. 2012) has been devised to measure the development of tissue-engineered constructs from human mesenchymal stem cells both *in vitro* before implantation (Xu et al. 2006) and *in vivo* after implantation in animal models (Othman et al. 2012). The potential of combining MRE with MR spectroscopy and other MRI parameters for tissue engineering is reviewed in Kotecha et al. (2013).

3.7 Future Challenges

A number of limitations and technical challenges remain before MRE can be adopted for the full gamut of prospective clinical applications. Although patient scan times should be minimized as far as possible, often full 3D coverage of an organ is required for accurate diagnosis, which implies a longer acquisition time. Innovations in MRE pulse sequences, such as fractionally encoded gradient echo sequences (Garteiser et al. 2013) and spatially selective excitations (Glaser et al. 2006) will allow shorter scan times for full 3D coverage.

For dynamic MRE in viscoelastic tissue, a trade-off exists between effective spatial resolution and coverage of the full tissue volume. The resolving power of MRE for features with varying elasticity is stronger at higher mechanical wave frequencies. However, in viscoelastic media, higher frequency waves are attenuated more rapidly, thereby limiting the distance from the driver for which adequate wave amplitudes may be measured. The issue of reduced coverage at higher frequencies might be offset by an improved driver technology, such as a phased-array acoustic driver (Mariappan et al. 2009). Furthermore, stiffer tissues such as bone and cartilage require higher mechanical wave frequencies for dynamic MRE (on the order of kHz), and clinical MRI scanner hardware often does not allow delivery of oscillating motion encoding gradients at higher frequencies. Hence, hardware improvements are necessary for applications to high elastic modulus tissue, such as implemented for *ex vivo* bovine cartilage in Lopez et al. (2008).

Another issue with regard to wave-delivery drivers is the potential sensitivity of the MRE-measured tissue biomechanics to the particular means of wave delivery, as predicted for brain MRE by the simulations in McGrath et al. (2016). Hence, future studies should compare wave-delivery methods for brain and other organ sites to determine the robustness and repeatability of measures, and to establish a consensus on the optimum wave-delivery methods. Sensitivity to other methodological variations such as pulse sequences, imaging systems, and magnetic field strengths must also be fully evaluated, such as in a recent study to determine the repeatability of liver stiffness measures (Trout et al. 2016). Another source of influence on MRE measures is the choice of inversion method employed, as some methods are prone to error, such as through amplification of imaging-related noise, or cannot handle the effects of wave interference. Further work is therefore required to determine the optimum inversion methods for different MRE applications, through comparison studies such as Yoshimitsu et al. (2016). Moreover, as biological tissue often has heterogeneous and anisotropic material properties, development of inversion methods based on more complex material constitutive models is required, such as those already explored for anisotropic (Sinkus et al. 2000, Romano et al. 2012), heterogeneous (Honarvar et al. 2013), and poroelastic material models (Perrinez et al. 2009).

3.8 Conclusion

MRE is an exciting emerging technology that has so far been demonstrated to be an important and powerful diagnostic tool for liver disease with the potential to become equally important for many other organ sites and diseases. MRE consists of two major classes of method: (1) dynamic and (2) quasi-static (or static). The dynamic methodologies have so far demonstrated the widest utility for *in vivo* clinical applications, whereas quasi-static methods have particular advantages for *in vitro* applications. An immense variety of MRE methodologies and clinical applications have been explored so far; however, further validation and optimization are required for many of these applications before they can be fully adopted into routine clinical practice.

References

An, H., Y. Shi, Q. Guo, and Y. Liu. 2016. Test-retest reliability of 3D EPI MR elastography of the pancreas. *Clin Radiol* 71 (10):1068.e7–1068.e12. doi:10.1016/j.crad.2016.03.014.

Anderson, A. T., E. E. Van Houten, M. D. McGarry, K. D. Paulsen, J. L. Holtrop, B. P. Sutton, J. G. Georgiadis, and C. L. Johnson. 2016. Observation of direction-dependent mechanical properties in the human brain with multi-excitation MR elastography. *J Mech Behav Biomed Mater* 59:538–546. doi:10.1016/j.jmbbm.2016.03.005.

Arani, A., M. Da Rosa, E. Ramsay, D. B. Plewes, M. A. Haider, and R. Chopra. 2013. Incorporating endorectal MR elastography into multi-parametric MRI for prostate cancer imaging: Initial feasibility in volunteers. *J Magn Reson Imaging* 38 (5):1251–1260. doi:10.1002/jmri.24028.

Arani, A., K. L. Glaser, S. P. Arunachalam, P. J. Rossman, D. S. Lake, J. D. Trzasko, A. Manduca, K. P. McGee, R. L. Ehman, and P. A. Araoz. 2016. In vivo, high-frequency three-dimensional cardiac MR elastography: Feasibility in normal volunteers. *Magn Reson Med* 77:351–360. doi:10.1002/mrm.26101.

Arani, A., M. C. Murphy, K. J. Glaser, A. Manduca, D. S. Lake, S. A. Kruse, C. R. Jack, Jr., R. L. Ehman, and J. Huston. 2015. Measuring the effects of aging and sex on regional brain stiffness with MR elastography in healthy older adults. *Neuroimage* 111:59–64. doi:10.1016/j.neuroimage.2015.02.016.

Asbach, P., D. Klatt, U. Hamhaber, J. Braun, R. Somasundaram, B. Hamm, and I. Sack. 2008. Assessment of liver viscoelasticity using multifrequency MR elastography. *Magn Reson Med* 60 (2):373–379. doi:10.1002/mrm.21636.

Axel, L., and L. Dougherty. 1989. MR imaging of motion with spatial modulation of magnetization. *Radiology* 171 (3):841–845. doi:10.1148/radiology.171.3.2717762.

Bahn, M. M., M. D. Brennan, R. S. Bahn, D. S. Dean, J. L. Kugel, and R. L. Ehman. 2009. Development and application of magnetic resonance elastography of the normal and pathological thyroid gland in vivo. *J Magn Reson Imaging* 30 (5):1151–1154. doi:10.1002/jmri.21963.

Barnhill, E., L. Hollis, I. Sack, J. Braun, P. R. Hoskins, P. Pankaj, C. Brown, E. J. van Beek, and N. Roberts. 2016. Nonlinear multiscale regularisation in MR elastography: Towards fine feature mapping. *Med Image Anal* 35:133–145. doi:10.1016/j.media.2016.05.012.

Basford, J. R., T. R. Jenkyn, K. N. An, R. L. Ehman, G. Heers, and K. R. Kaufman. 2002. Evaluation of healthy and diseased muscle with magnetic resonance elastography. *Arch Phys Med Rehabil* 83 (11):1530–1536.

Bensamoun, S. F., S. I. Ringleb, Q. Chen, R. L. Ehman, K. N. An, and M. Brennan. 2007. Thigh muscle stiffness assessed with magnetic resonance elastography in hyperthyroid patients before and after medical treatment. *J Magn Reson Imaging* 26 (3):708–713. doi:10.1002/jmri.21073.

Bishop, J., A. Samani, J. Sciarretta, and D. B. Plewes. 2000. Two-dimensional MR elastography with linear inversion reconstruction: Methodology and noise analysis. *Phys Med Biol* 45:2081–2091.

Boulet, T., M. L. Kelso, and S. F. Othman. 2011. Microscopic magnetic resonance elastography of traumatic brain injury model. *J Neurosci Methods* 201 (2):296–306. doi:10.1016/j.jneumeth.2011.08.019.

Brauck, K., C. J. Galban, S. Maderwald, B. L. Herrmann, and M. E. Ladd. 2007. Changes in calf muscle elasticity in hypogonadal males before and after testosterone substitution as monitored by magnetic resonance elastography. *Eur J Endocrinol* 156 (6):673–678. doi:10.1530/eje-06-0694.

Braun, J., K. Braun, and I. Sack. 2003. Electromagnetic actuator for generating variably oriented shear waves in MR elastography. *Magn Reson Med* 50 (1):220–222. doi:10.1002/mrm.10479.

Braun, J., J. Guo, R. Lutzkendorf, J. Stadler, S. Papazoglou, S. Hirsch, I. Sack, and J. Bernarding. 2014. High-resolution mechanical imaging of the human brain by three-dimensional multifrequency magnetic resonance elastography at 7T. *Neuroimage* 90:308–314. doi:10.1016/j.neuroimage.2013.12.032.

Brown, R. W., Y.-C. N. Cheng, E. M. Haacke, M. R. Thompson, and R. Venkatesan. 2014. *Magnetic Resonance Imaging: Physical Principles and Sequence Design.* Hoboken, NJ: John Wiley & Sons.

Catheline, S., F. Wu, and M. Fink. 1999. A solution to diffraction biases in sonoelasticity: the acoustic impulse technique. *J Acoust Soc Am* 105 (5):2941–2950.

Chen, J., C. Ni, and T. Zhuang. 2005. Mechanical shear wave induced by piezoelectric ceramics for magnetic resonance elastography. *Conf Proc IEEE Eng Med Biol Soc* 7:7020–7023. doi:10.1109/iembs.2005.1616122.

Chen, J., J. A. Talwalkar, M. Yin, K. J. Glaser, S. O. Sanderson, and R. L. Ehman. 2011. Early detection of nonalcoholic steatohepatitis in patients with nonalcoholic fatty liver disease by using MR elastography. *Radiology* 259 (3):749–756. doi:10.1148/radiol.11101942.

Chenevert, T. L., A. R. Skovoroda, M. O'Donnell, and S. Y. Emelianov. 1998. Elasticity reconstructive imaging by means of stimulated echo MRI. *Magn Reson Med* 39:482–490.

Chopra, R., A. Arani, Y. Huang, M. Musquera, J. Wachsmuth, M. Bronskill, and D. Plewes. 2009. In vivo MR elastography of the prostate gland using a transurethral actuator. *Magn Reson Med* 62:665–671.

Curtis, E. T., S. Zhang, V. Khalilzad-Sharghi, T. Boulet, and S. F. Othman. 2012. Magnetic resonance elastography methodology for the evaluation of tissue engineered construct growth. *J Vis Exp* (60). doi:10.3791/3618.

Damughatla, A. R., B. Raterman, T. Sharkey-Toppen, N. Jin, O. P. Simonetti, R. D. White, and A. Kolipaka. 2015. Quantification of aortic stiffness using MR elastography and its comparison to MRI-based pulse wave velocity. *J Magn Reson Imaging* 41 (1):44–51. doi:10.1002/jmri.24506.

Debernard, L., L. Robert, F. Charleux, and S. F. Bensamoun. 2011. Analysis of thigh muscle stiffness from childhood to adulthood using magnetic resonance elastography (MRE) technique. *Clin Biomech (Bristol, Avon)* 26 (8):836–840. doi:10.1016/j.clinbiomech.2011.04.004.

Debernard, L., L. Robert, F. Charleux, and S. F. Bensamoun. 2013. A possible clinical tool to depict muscle elasticity mapping using magnetic resonance elastography. *Muscle Nerve* 47 (6):903–908. doi:10.1002/mus.23678.

Di Ieva, A., F. Grizzi, E. Rognone, Z. T. Tse, T. Parittotokkaporn, Y. Baena F. Rodriguez et al. 2010. Magnetic resonance elastography: A general overview of its current and future applications in brain imaging. *Neurosurg Rev* 33 (2):137–145. doi:10.1007/s10143-010-0249-6.

Dittmann, F., H. Tzschatzsch, S. Hirsch, E. Barnhill, J. Braun, I. Sack, and J. Guo. 2016. Tomoelastography of the abdomen: Tissue mechanical properties of the liver, spleen, kidney, and pancreas from single MR elastography scans at different hydration states. *Magn Reson Med* 78:976–983. doi:10.1002/mrm.26484.

Domire, Z. J., M. B. McCullough, Q. Chen, and K. N. An. 2009. Feasibility of using magnetic resonance elastography to study the effect of aging on shear modulus of skeletal muscle. *J Appl Biomech* 25 (1):93–97.

Doyley, M. M. 2012. Model-based elastography: A survey of approaches to the inverse elasticity problem. *Phys Med Biol* 57 (3):R35–R73. doi: 10.1088/0031-9155/57/3/r35.

Ehman, E. C., P. J. Rossman, S. A. Kruse, A. V. Sahakian, and K. J. Glaser. 2008. Vibration safety limits for magnetic resonance elastography. *Phys Med Biol* 53 (4):925–935. doi:10.1088/0031-9155/53/4/007.

Elgeti, T., M. Laule, N. Kaufels, J. Schnorr, B. Hamm, A. Samani, J. Braun, and I. Sack. 2009. Cardiac MR elastography: Comparison with left ventricular pressure measurement. *J Cardiovasc Magn Reson* 11:44. doi:10.1186/1532-429x-11-44.

Etchell, E., L. Juge, A. Hatt, R. Sinkus, and L. E. Bilston. 2016. Liver stiffness values are lower in pediatric subjects than in adults and increase with age: A multifrequency MR elastography study. *Radiology* 283:222–230. doi:10.1148/radiol.2016160252.

Fattahi, N., A. Arani, A. Perry, F. Meyer, A. Manduca, K. Glaser, M. L. Senjem, R. L. Ehman, and J. Huston. 2016. MR elastography demonstrates increased brain stiffness in normal pressure hydrocephalus. *AJNR Am J Neuroradiol* 37 (3):462–467. doi:10.3174/ajnr.A4560.

Fehlner, A., S. Papazoglou, M. D. McGarry, K. D. Paulsen, J. Guo, K. J. Streitberger, S. Hirsch, J. Braun, and I. Sack. 2015. Cerebral multifrequency MR elastography by remote excitation of intracranial shear waves. *NMR Biomed* 28:1426–1432.

Feng, Y., E. H. Clayton, Y. Chang, R. J. Okamoto, and P. V. Bayly. 2013. Viscoelastic properties of the ferret brain measured in vivo at multiple frequencies by magnetic resonance elastography. *J Biomech* 46 (5):863–870. doi:10.1016/j.jbiomech.2012.12.024.

Freimann, F. B., S. Muller, K. J. Streitberger, J. Guo, S. Rot, A. Ghori, P. Vajkoczy, R. Reiter, I. Sack, and J. Braun. 2013. MR elastography in a murine stroke model reveals correlation of macroscopic viscoelastic properties of the brain with neuronal density. *NMR Biomed* 26 (11):1534–1539. doi:10.1002/nbm.2987.

Freimann, F. B., K. J. Streitberger, D. Klatt, K. Lin, J. McLaughlin, J. Braun, C. Sprung, and I. Sack. 2012. Alteration of brain viscoelasticity after shunt treatment in normal pressure hydrocephalus. *Neuroradiology* 54 (3):189–196. doi:10.1007/s00234-011-0871-1.

Gallichan, D., M. D. Robson, A. Bartsch, and K. L. Miller. 2009. TREMR: Table-resonance elastography with MR. *Magn Reson Med* 62 (3):815–821. doi:10.1002/mrm.22046.

Garteiser, P., R. S. Sahebjavaher, L. C. Ter Beek, S. Salcudean, V. Vilgrain, B. E. Van Beers, and R. Sinkus. 2013. Rapid acquisition of multifrequency, multislice and multidirectional MR elastography data with a fractionally encoded gradient echo sequence. *NMR Biomed* 26 (10):1326–1335. doi:10.1002/nbm.2958.

Glaser, K. J., J. P. Felmlee, and R. L. Ehman. 2006. Rapid MR elastography using selective excitations. *Magn Reson Med* 55 (6):1381–1389. doi:10.1002/mrm.20913.

Godfrey, E. M., A. J. Patterson, A. N. Priest, S. E. Davies, I. Joubert, A. S. Krishnan, N. Griffin et al. 2012. A comparison of MR elastography and 31P MR spectroscopy with histological staging of liver fibrosis. *Eur Radiol* 22 (12):2790–2797. doi:10.1007/s00330-012-2527-x.

Goss, S. A., R. L. Johnston, and F. Dunn. 1978. Comprehensive compilation of empirical ultrasonic properties of mammalian tissues. *J Acoust Soc Am* 64 (2):423–457.

Green, M. A., L. E. Bilston, and R. Sinkus. 2008. In vivo brain viscoelastic properties measured by magnetic resonance elastography. *NMR Biomed* 21 (7):755–764. doi:10.1002/nbm.1254.

Guo, J., C. Buning, E. Schott, T. Kroncke, J. Braun, I. Sack, and C. Althoff. 2015. In vivo abdominal magnetic resonance elastography for the assessment of portal hypertension before and after transjugular intrahepatic portosystemic shunt implantation. *Invest Radiol* 50 (5):347–351. doi:10.1097/rli.0000000000000136.

Guo, J., S. Hirsch, A. Fehlner, S. Papazoglou, M. Scheel, J. Braun, and I. Sack. 2013. Towards an elastographic atlas of brain anatomy. *PLoS One* 8 (8):e71807. doi:10.1371/journal.pone.0071807.

Hardy, P. A., A. C. Ridler, C. B. Chiarot, D. B. Plewes, and R. M. Henkelman. 2005. Imaging articular cartilage under compression: Cartilage elastography. *Magn Reson Med* 53 (5):1065–1073. doi:10.1002/mrm.20439.

Hawley, J. R., P. Kalra, X. Mo, B. Raterman, L. D. Yee, and A. Kolipaka. 2016. Quantification of breast stiffness using MR elastography at 3 Tesla with a soft sternal driver: A reproducibility study. *J Magn Reson Imaging* 45:1379–1384. doi:10.1002/jmri.25511.

Heers, G., T. Jenkyn, M. A. Dresner, M. O. Klein, J. R. Basford, K. R. Kaufman, R. L. Ehman, and K. N. An. 2003. Measurement of muscle activity with magnetic resonance elastography. *Clin Biomech (Bristol, Avon)* 18 (6):537–542.

Hennedige, T. P., J. T. Hallinan, F. P. Leung, L. L. Teo, S. Iyer, G. Wang, S. Chang, K. K. Madhavan, A. Wee, and S. K. Venkatesh. 2016. Comparison of magnetic resonance elastography and diffusion-weighted imaging for differentiating benign and malignant liver lesions. *Eur Radiol* 26 (2):398–406. doi:10.1007/s00330-015-3835-8.

Hirsch, S., D. Klatt, F. Freimann, M. Scheel, J. Braun, and I. Sack. 2013. In vivo measurement of volumetric strain in the human brain induced by arterial pulsation and harmonic waves. *Magn Reson Med* 70 (3):671–683. doi:10.1002/mrm.24499.

Honarvar, M., R. Sahebjavaher, R. Sinkus, R. Rohling, and S. Salcudean. 2013. Curl-based finite element reconstruction of the shear modulus without assuming local homogeneity: Time harmonic case. *IEEE Trans Med Imaging* 32:2189–2199. doi:10.1109/tmi.2013.2276060.

Hong, S. H., S. J. Hong, J. S. Yoon, C. H. Oh, J. G. Cha, H. K. Kim, and B. Bolster, Jr. 2016. Magnetic resonance elastography (MRE) for measurement of muscle stiffness of the shoulder: Feasibility with a 3 T MRI system. *Acta Radiol* 57 (9):1099–1106. doi:10.1177/0284185115571987.

Huwart, L., F. Peeters, R. Sinkus, L. Annet, N. Salameh, L. C. ter Beek, Y. Horsmans, and B. E. Van Beers. 2006. Liver fibrosis: non-invasive assessment with MR elastography. *NMR Biomed* 19:173–179.

Huwart, L., C. Sempoux, N. Salameh, J. Jamart, L. Annet, R. Sinkus, F. Peeters, L. C. ter Beek, Y. Horsmans, and B. E. Van Beers. 2007. Liver fibrosis: Noninvasive assessment with MR elastography versus aspartate aminotransferase-to-platelet ratio index. *Radiology* 245 (2):458–466.

Huwart, L., C. Sempoux, E. Vicaut, N. Salameh, L. Annet, E. Danse, F. Peeters et al. 2008. Magnetic resonance elastography for the noninvasive staging of liver fibrosis. *Gastroenterology* 135 (1):32–40. doi:10.1053/j.gastro.2008.03.076.

Ito, D., T. Numano, K. Mizuhara, K. Takamoto, T. Onishi, and H. Nishijo. 2016. A new technique for MR elastography of the supraspinatus muscle: A gradient-echo type multi-echo sequence. *Magn Reson Imaging* 34 (8):1181–1188. doi:10.1016/j.mri.2016.06.003.

Itoh, Y., Y. Takehara, T. Kawase, K. Terashima, Y. Ohkawa, Y. Hirose, A. Koda et al. 2016. Feasibility of magnetic resonance elastography for the pancreas at 3T. *J Magn Reson Imaging* 43 (2):384–390. doi:10.1002/jmri.24995.

Jenkyn, T. R., R. L. Ehman, and K. N. An. 2003. Noninvasive muscle tension measurement using the novel technique of magnetic resonance elastography (MRE). *J Biomech* 36 (12):1917–1921.

Jiang, X., P. Asbach, K. J. Streitberger, A. Thomas, B. Hamm, J. Braun, I. Sack, and J. Guo. 2014. In vivo high-resolution magnetic resonance elastography of the uterine corpus and cervix. *Eur Radiol* 24 (12):3025–3033. doi:10.1007/s00330-014-3305-8.

Kemper, J., R. Sinkus, J. Lorenzen, C. Nolte-Ernsting, A. Stork, and G. Adam. 2004. MR elastography of the prostate: Initial in-vivo application. *Rofo. Fortschritte auf dem Gebiet der Rontgenstrahlen und der bildgebenden Verfahren* 176 (8):1094–1099.

Kenyhercz, W. E., B. Raterman, V. S. Illapani, J. Dowell, X. Mo, R. D. White, and A. Kolipaka. 2016. Quantification of aortic stiffness using magnetic resonance elastography: Measurement reproducibility, pulse wave velocity comparison, changes over cardiac cycle, and relationship with age. *Magn Reson Med* 75 (5):1920–1926. doi:10.1002/mrm.25719.

Klatt, D., U. Hamhaber, P. Asbach, J. Braun, and I. Sack. 2007. Noninvasive assessment of the rheological behavior of human organs using multifrequency MR elastography: A study of brain and liver viscoelasticity. *Phys Med Biol* 52:7281–7294.

Kolipaka, A., V. S. Illapani, P. Kalra, J. Garcia, X. Mo, M. Markl, and R. D. White. 2016. Quantification and comparison of 4D-flow MRI-derived wall shear stress and MRE-derived wall stiffness of the abdominal aorta. *J Magn Reson Imaging* 45:771–778. doi:10.1002/jmri.25445.

Kolipaka, A., D. Woodrum, P. A. Araoz, and R. L. Ehman. 2012. MR elastography of the in vivo abdominal aorta: A feasibility study for comparing aortic stiffness between hypertensives and normotensives. *J Magn Reson Imaging* 35 (3):582–586. doi:10.1002/jmri.22866.

Kotecha, M., D. Klatt, and R. L. Magin. 2013. Monitoring cartilage tissue engineering using magnetic resonance spectroscopy, imaging, and elastography. *Tissue Eng Part B Rev* 19 (6):470–484. doi:10.1089/ten.TEB.2012.0755.

Lee, C. U., J. F. Glockner, K. J. Glaser, M. Yin, J. Chen, A. Kawashima, B. Kim, W. K. Kremers, R. L. Ehman, and J. M. Gloor. 2012. MR elastography in renal transplant patients and correlation with renal allograft biopsy: A feasibility study. *Acad Radiol* 19 (7):834–841. doi:10.1016/j.acra.2012.03.003.

Leitao, H. S., S. Doblas, P. Garteiser, G. d'Assignies, V. Paradis, F. Mouri, C. F. Geraldes, M. Ronot, and B. E. Van Beers. 2016. Hepatic fibrosis, inflammation, and steatosis: Influence on the MR viscoelastic and diffusion parameters in patients with chronic liver disease. *Radiology* 283:98–107. doi:10.1148/radiol.2016151570.

Li, S., M. Chen, W. Wang, W. Zhao, J. Wang, X. Zhao, and C. Zhou. 2011. A feasibility study of MR elastography in the diagnosis of prostate cancer at 3.0T. *Acta Radiol* 52 (3):354–358. doi:10.1258/ar.2010.100276.

Lipp, A., R. Trbojevic, F. Paul, A. Fehlner, S. Hirsch, M. Scheel, C. Noack, J. Braun, and I. Sack. 2013. Cerebral magnetic resonance elastography in supranuclear palsy and idiopathic Parkinson's disease. *Neuroimage Clin* 3:381–387. doi:10.1016/j.nicl.2013.09.006.

Litwiller, D. V., S. J. Lee, A. Kolipaka, Y. K. Mariappan, K. J. Glaser, J. S. Pulido, and R. L. Ehman. 2010. Magnetic resonance elastography of the ex-vivo bovine globe. *J Magn Reson Imaging* 32 (1):44–51.

Lopez, O., K. K. Amrami, A. Manduca, and R. L. Ehman. 2008. Characterization of the dynamic shear properties of hyaline cartilage using high frequency dynamic MR elastography. *Magn Reson Med* 59:356–364.

Lorenzen, J., R. Sinkus, M. Lorenzen, M. Dargatz, C. Leussler, P. Roschmann, and G. Adam. 2002. MR elastography of the breast: Preliminary clinical results. *Rofo* 174 (7):830–834. doi:10.1055/s-2002-32690.

Low, G., N. E. Owen, I. Joubert, A. J. Patterson, M. J. Graves, G. J. Alexander, and D. J. Lomas. 2015a. Magnetic resonance elastography in the detection of hepatorenal syndrome in patients with cirrhosis and ascites. *Eur Radiol* 25 (10):2851–2858. doi:10.1007/s00330-015-3723-2.

Low, G., N. E. Owen, I. Joubert, A. J. Patterson, M. J. Graves, K. J. Glaser, G. J. Alexander, and D. J. Lomas. 2015b. Reliability of magnetic resonance elastography using multislice two-dimensional spin-echo echo-planar imaging (SE-EPI) and three-dimensional inversion reconstruction for assessing renal stiffness. *J Magn Reson Imaging* 42 (3):844–850. doi:10.1002/jmri.24826.

Manduca, A., T. E. Oliphant, M. A. Dresner, J. L. Mahowald, S. A. Kruse, E. Amromin, J. P. Felmlee, J. F. Greenleaf, and R. L. Ehman. 2001. Magnetic resonance elastography: Non-invasive mapping of tissue elasticity. *Med Image Anal* 5:237–254.

Mariappan, Y. K., K. J. Glaser, R. D. Hubmayr, A. Manduca, R. L. Ehman, and K. P. McGee. 2011. MR elastography of human lung parenchyma: Technical development, theoretical modeling and in vivo validation. *J Magn Reson Imaging* 33 (6):1351–1361. doi:10.1002/jmri.22550.

Mariappan, Y. K., K. J. Glaser, D. L. Levin, R. Vassallo, R. D. Hubmayr, C. Mottram, R. L. Ehman, and K. P. McGee. 2014. Estimation of the absolute shear stiffness of human lung parenchyma using (1) H spin echo, echo planar MR elastography. *J Magn Reson Imaging* 40 (5):1230–1237. doi:10.1002/jmri.24479.

Mariappan, Y. K., P. J. Rossman, K. J. Glaser, A. Manduca, and R. L. Ehman. 2009. Magnetic resonance elastography with a phased-array acoustic driver system. *Magn Reson Med* 61 (3):678–685. doi:10.1002/mrm.21885.

Mariappan, Y. K., K. J. Glaser, and R. L. Ehman. 2010. Magnetic resonance elastography: A review. *Clin Anat* 23:497–511.

Mazumder, R., S. Schroeder, X. Mo, B. D. Clymer, R. D. White, and A. Kolipaka. 2016. In vivo quantification of myocardial stiffness in hypertensive porcine hearts using MR elastography. *J Magn Reson Imaging* 45:813–820. doi:10.1002/jmri.25423.

McCracken, P. J., A. Manduca, J. Felmlee, and R. L. Ehman. 2005. Mechanical transient-based magnetic resonance elastography. *Magn Reson Med* 53 (3):628–639. doi:10.1002/mrm.20388.

McCullough, M. B., Z. J. Domire, A. M. Reed, S. Amin, S. R. Ytterberg, Q. Chen, and K. N. An. 2011. Evaluation of muscles affected by myositis using magnetic resonance elastography. *Muscle Nerve* 43 (4):585–590. doi:10.1002/mus.21923.

McGee, K. P., R. D. Hubmayr, and R. L. Ehman. 2008. MR elastography of the lung with hyperpolarized 3He. *Magn Reson Med* 59:14–18.

McGee, K. P., Y. K. Mariappan, R. D. Hubmayr, R. E. Carter, Z. Bao, D. L. Levin, A. Manduca, and R. L. Ehman. 2012. Magnetic resonance assessment of parenchymal elasticity in normal and edematous, ventilator-injured lung. *J Appl Physiol* 113 (4):666–676. doi:10.1152/japplphysiol.01628.2011.

McGrath, D. M., W. D. Foltz, A. Al-Mayah, C. J. Niu, and K. K. Brock. 2012. Quasi-static magnetic resonance elastography at 7 T to measure the effect of pathology before and after fixation on tissue biomechanical properties. *Magn Reson Med* 68 (1):152–165. doi:10.1002/mrm.23223.

McGrath, D. M., J. Lee, W. D. Foltz, N. Samavati, T. van der Kwast, M. A. Jewett, P. Chung, C. Menard, and K. K. Brock. 2017. MR elastography to measure the effects of cancer and pathology fixation on prostate biomechanics, and comparison with T 1, T 2 and ADC. *Phys Med Biol* 62 (3):1126–1148. doi:10.1088/1361-6560/aa52f4.

McGrath, D. M., N. Ravikumar, L. Beltrachini, I. D. Wilkinson, A. F. Frangi, and Z. A. Taylor. 2016. Evaluation of wave delivery methodology for brain MRE: Insights from computational simulations. *Magn Reson Med* 78:341–356. doi:10.1002/mrm.26333.

McGrath, D. M., N. Ravikumar, I. D. Wilkinson, A. F. Frangi, and Z. Taylor. 2015. Magnetic resonance elastography of the brain: An in silico study to determine the influence of cranial anatomy. *Magn Reson Med* 76:645–662. doi:10.1002/mrm.25881.

McGrath, D. M., W. D. Foltz, N. Samavati, J. Lee, M. A. Jewett, T. H. van der Kwast, C. Menard, and K. K. Brock. 2011. Biomechanical property quantification of prostate cancer by quasi-static MR elastography at 7 tesla of radical prostatectomy, and correlation with whole mount histology. *Proc Intl Soc Mag Res Med* 19:1483.

McKnight, A. L., J. L. Kuge, P. J. Rossman, A. Manduca, L. C. Hartman, and R. L. Ehman. 2002. MR elastography of breast cancer: Preliminary results. *Amer J Roent* 178:1411–1417.

McRobbie, D. W., E. A. Moore, M. J. Graves, and M. R. Prince. 2007. *MRI from Picture to Proton*. Cambridge, UK: Cambridge University Press.

Murphy, M. C., D. T. Jones, C. R. Jack, Jr., K. J. Glaser, M. L. Senjem, A. Manduca, J. P. Felmlee, R. E. Carter, R. L. Ehman, and J. Huston. 2016. Regional brain stiffness changes across the Alzheimer's disease spectrum. *Neuroimage Clin* 10:283–290. doi:10.1016/j.nicl.2015.12.007.

Murphy, M. C., J. III Huston, C. R. Jr Jack, K. J. Glaser, A. Manduca, J. P. Felmlee, and R. L. Ehman. 2011. Decreased brain stiffness in Alzheimer's disease determined by magnetic resonance elastography. *J Magn Reson Im* 34:494–498.

Muthupillai, R., D. J. Lomas, P. J. Rossman, J. F. Greenleaf, A. Manduca, and R. L. Ehman. 1995. Magnetic resonance elastography by direct visualisation of propagating acoustic strain waves. *Science* 269 (5232):1854–1857.

O'Donnell, M. 1985. NMR blood flow imaging using multiecho, phase contrast sequences. *Med Phys* 12 (1):59–64. doi:10.1118/1.595736.

Oliphant, T. E., A. Manduca, R. L. Ehman, and J. F. Greenleaf. 2001. Complex-valued stiffness reconstruction for magnetic resonance elastography by algebraic inversion of the differential equation. *Magn Reson Med* 45 (2):299–310.

Osman, N. F. 2003. Detecting stiff masses using strain-encoded (SENC) imaging. *Magn Reson Med* 49 (3):605–608. doi:10.1002/mrm.10376.

Othman, S. F., E. T. Curtis, S. A. Plautz, A. K. Pannier, S. D. Butler, and H. Xu. 2012. MR elastography monitoring of tissue-engineered constructs. *NMR Biomed* 25 (3):452–463. doi:10.1002/nbm.1663.

Othman, S. F., H. Xu, T. J. Royston, and R. L. Magin. 2005. Microscopic magnetic resonance elastography (microMRE). *Magn Reson Med* 54 (3):605–615. doi:10.1002/mrm.20584.

Perrinez, P. R., F. E. Kennedy, E. E. Van Houten, J. B. Weaver, and K. D. Paulsen. 2010. Magnetic resonance poroelastography: An algorithm for estimating the mechanical properties of fluid-saturated soft tissues. *IEEE Trans Med Imaging* 29 (3):746–755. doi:10.1109/tmi.2009.2035309.

Perrinez, P. R., F. E. Kennedy, E. W. Van Houten, J. B. Weaver, and K. D. Paulsen. 2009. Modeling of soft poroelastic tissue in time-harmonic MR elastography. *IEEE Trans Biomed Eng* 56 (3):598–608.

Perumpail, R. B., J. Levitsky, Y. Wang, V. S. Lee, J. Karp, N. Jin, G. Y. Yang et al. 2012. MRI-guided biopsy to correlate tissue specimens with MR elastography stiffness readings in liver transplants. *Acad Radiol* 19 (9):1121–1126. doi:10.1016/j.acra.2012.05.011.

Plewes, D. B., I. Betty, S. N. Urchuk, and I. Soutar. 1995. Visualizing tissue compliance with MR imaging. *J Magn Reson Imaging* 5 (6):733–738.

Plewes, D. B., J. Bishop, A. Samani, and J. Sciarretta. 2000. Visualization and quantification of breast cancer biomechanical properties with magnetic resonance elastography. *Phys Med Biol* 45:1591–1610.

Pong, A. C., L. Juge, S. Cheng, and L. E. Bilston. 2016. Longitudinal measurements of postnatal rat brain mechanical properties in-vivo. *J Biomech* 49 (9):1751–1756. doi:10.1016/j.jbiomech.2016.04.005.

Reiss-Zimmermann, M., K. J. Streitberger, I. Sack, J. Braun, F. Arlt, D. Fritzsch, and K. T. Hoffmann. 2014. High resolution imaging of viscoelastic properties of intracranial tumours by multi-frequency magnetic resonance elastography. *Clin Neuroradiol* 25:371–378. doi:10.1007/s00062-014-0311-9.

Ringleb, S. I., S. F. Bensamoun, Q. Chen, A. Manduca, K. N. An, and R. L. Ehman. 2007. Applications of magnetic resonance elastography to healthy and pathologic skeletal muscle. *J Magn Reson Imaging* 25 (2):301–309. doi:10.1002/jmri.20817.

Robert, B., R. Sinkus, J. L. Gennisson, and M. Fink. 2009. Application of DENSE-MR-elastography to the human heart. *Magn Reson Med* 62 (5):1155–1163. doi:10.1002/mrm.22124.

Romano, A., J. Guo, T. Prokscha, T. Meyer, S. Hirsch, J. Braun, I. Sack, and M. Scheel. 2014. In vivo waveguide elastography: Effects of neurodegeneration in patients with amyotrophic lateral sclerosis. *Magn Reson Med* 72 (6):1755–1761. doi:10.1002/mrm.25067.

Romano, A. J., J. J. Shirron, and J. A. Bucaro. 1998. On the noninvasive determination of material parameters from a knowledge of elastic displacements theory and numerical simulation. *IEEE Trans Ultrason Ferroelectr Freq Control* 45 (3):751–759. doi:10.1109/58.677725.

Romano, A., M. Scheel, S. Hirsch, J. Braun, and I. Sack. 2012. In vivo waveguide elastography of white matter tracts in the human brain. *Magn Reson Med* 68 (5):1410–1422. doi:10.1002/mrm.24141.

Rouviere, O., R. Souchon, G. Pagnoux, J. M. Menager, and J. Y. Chapelon. 2011. Magnetic resonance elastography of the kidneys: Feasibility and reproducibility in young healthy adults. *J Magn Reson Imaging* 34 (4):880–886. doi:10.1002/jmri.22670.

Rouviere, O., M. Yin, M. A. Dresner, P. J. Rossman, L. J. Burgart, J. L. Fidler, and R. L. Ehman. 2006. MR Elastography of the liver: Preliminary results. *Radiology* 240 (2):440–448.

Sack, I., B. Beierbach, U. Hamhaber, D. Klatt, and J. Braun. 2008. Non-invasive measurement of brain viscoelasticity using magnetic resonance elastography. *NMR Biomed* 21 (3):265–271. doi:10.1002/nbm.1189.

Sack, I., B. Beierbach, J. Wuerfel, D. Klatt, U. Hamhaber, S. Papazoglou, P. Martus, and J. Braun. 2009. The impact of aging and gender on brain viscoelasticity. *Neuroimage* 46 (3):652–657. doi:10.1016/j.neuroimage.2009.02.040.

Sack, I., J. Bernarding, and J. Braun. 2002. Analysis of wave patterns in MR elastography of skeletal muscle using coupled harmonic oscillator simulations. *Magn Reson Imaging* 20 (1):95–104.

Sack, I., J. Rump, T. Elgeti, A. Samani, and J. Braun. 2009. MR elastography of the human heart: noninvasive assessment of myocardial elasticity changes by shear wave amplitude variations. *Magn Reson Med* 61 (3):668–677. doi:10.1002/mrm.21878.

Sahebjavaher, R. S., A. Baghani, M. Honarvar, R. Sinkus, and S. E. Salcudean. 2013. Transperineal prostate MR elastography: Initial in vivo results. *Magn Reson Med* 69 (2):411–420. doi:10.1002/mrm.24268.

Sahebjavaher, R. S., S. Frew, A. Bylinskii, L. ter Beek, P. Garteiser, M. Honarvar, R. Sinkus, and S. Salcudean. 2014. Prostate MR elastography with transperineal electromagnetic actuation and a fast fractionally encoded steady-state gradient echo sequence. *NMR Biomed* 27 (7):784–794. doi:10.1002/nbm.3118.

Sahebjavaher, R. S., G. Nir, L. O. Gagnon, J. Ischia, E. C. Jones, S. D. Chang, A. Yung et al. 2015. MR elastography and diffusion-weighted imaging of ex vivo prostate cancer: Quantitative comparison to histopathology. *NMR Biomed* 28 (1):89–100. doi:10.1002/nbm.3203.

Sakai, N., Y. Takehara, S. Yamashita, N. Ohishi, H. Kawaji, T. Sameshima, S. Baba, H. Sakahara, and H. Namba. 2016. Shear stiffness of 4 common intracranial tumors measured using MR elastography: Comparison with intraoperative consistency grading. *AJNR Am J Neuroradiol*. doi:10.3174/ajnr.A4832.

Samani, A., J. Bishop, C. Luginbuhl, and D. B. Plewes. 2003. Measuring the elastic modulus of ex vivo small tissue samples. *Phys Med Biol* 48:2183–2198.

Samani, A., J. Bishop, and D. B. Plewes. 2001. A constrained modulus reconstruction technique for breast cancer assessment. *IEEE Trans Med Imaging* 20 (9):877–885.

Sandrin, L., B. Fourquet, J. M. Hasquenoph, S. Yon, C. Fournier, F. Mal, C. Christidis et al. 2003. Transient elastography: A new noninvasive method for assessment of hepatic fibrosis. *Ultrasound Med Biol* 29 (12):1705–1713.

Sarvazyan, A. 1998. Mechanical imaging: A new technology for medical diagnostics. *Int J Med Inform* 49 (2):195–216.

Sarvazyan, A., T. J. Hall, M. W. Urban, M. Fatemi, S. R. Aglyamov, and B. S. Garra. 2011. An overview of elastography: An emerging branch of medical imaging. *Curr Med Imaging Rev* 7 (4):255–282.

Sarvazyan, A. P., O. V. Rudenko, S. D. Swanson, J. B. Fowlkes, and S. Y. Emelianov. 1998. Shear wave elasticity imaging: A new ultrasonic technology of medical diagnostics. *Ultrasound Med Biol* 24 (9):1419–1435.

Sarvazyan, A. P., A. R. Skovoroda, S. Y. Emelianov, B. J. Fowlkes, J. G. Pipe, R. S. Adler, R. B. Buxton, and P. L. Carson. 1995. Biophysical bases of elasticity imaging. In *Acoustical Imaging*, J. P. Jones (Ed.), 223–240. New York: Plenum Press.

Schregel, K., E. Wuerfel, P. Garteiser, I. Gemeinhardt, T. Prozorovski, O. Aktas, H. Merz, D. Petersen, J. Wuerfel, and R. Sinkus. 2012. Demyelination reduces brain parenchymal stiffness quantified in vivo by magnetic resonance elastography. *Proc Natl Acad Sci USA* 109 (17):6650–6655.

Shi, Y., K. J. Glaser, S. K. Venkatesh, E. I. Ben-Abraham, and R. L. Ehman. 2015. Feasibility of using 3D MR elastography to determine pancreatic stiffness in healthy volunteers. *J Magn Reson Imaging* 41 (2):369–375. doi:10.1002/jmri.24572.

Shin, S. U., J. M. Lee, M. H. Yu, J. H. Yoon, J. K. Han, B. I. Choi, K. J. Glaser, and R. L. Ehman. 2014. Prediction of esophageal varices in patients with cirrhosis: Usefulness of three-dimensional MR elastography with echo-planar imaging technique. *Radiology* 272 (1):143–153. doi:10.1148/radiol.14130916.

Siegmann, K. C., T. Xydeas, R. Sinkus, B. Kraemer, U. Vogel, and C. D. Claussen. 2010. Diagnostic value of MR elastography in addition to contrast-enhanced MR imaging of the breast-initial clinical results. *Eur Radiol* 20 (2):318–325. doi:10.1007/s00330-009-1566-4.

Singh, S., S. K. Venkatesh, Z. Wang, F. H. Miller, U. Motosugi, R. N. Low, T. Hassanein et al. 2015. Diagnostic performance of magnetic resonance elastography in staging liver fibrosis: A systematic review and meta-analysis of individual participant data. *Clin Gastroenterol Hepatol* 13 (3):440–451.e6. doi:10.1016/j.cgh.2014.09.046.

Sinkus, R., J. Lorenzen, D. Schrader, M. Lorenzen, M. Dargatz, and D. Holz. 2000. High-resolution tensor MR elastography for breast tumour detection. *Phys Med Biol* 45 (6):1649–1664.

Sinkus, R., K. Siegmann, T. Xydeas, M. Tanter, C. Claussen, and M. Fink. 2007. MR elastography of breast lesions: Understanding the solid/liquid duality can improve the specificity of contrast-enhanced MR mammography. *Magn Reson Med* 58 (6):1135–1144. doi:10.1002/mrm.21404.

Sinkus, R., M. Tanter, T. Xydeas, S. Catheline, J. Bercoff, and M. Fink. 2005. Viscoelastic shear properties of in vivo breast lesions measured by MR elastography. *Magn Reson Imaging* 23 (2):159–165. doi:10.1016/j.mri.2004.11.060.

Souchon, R., R. Salomir, O. Beuf, L. Milot, D. Grenier, D. Lyonnet, J. Y. Chapelon, and O. Rouviere. 2008. Transient MR elastography (t-MRE) using ultrasound radiation force: Theory, safety, and initial experiments in vitro. *Magn Reson Med* 60 (4):871–881. doi:10.1002/mrm.21718.

Streitberger, K. J., M. Reiss-Zimmermann, F. B. Freimann, S. Bayerl, J. Guo, F. Arlt, J. Wuerfel, J. Braun, K. T. Hoffmann, and I. Sack. 2014. High-resolution mechanical imaging of glioblastoma by multifrequency magnetic resonance elastography. *PLoS One* 9 (10):e110588. doi:10.1371/journal.pone.0110588.

Streitberger, K. J., I. Sack, D. Krefting, C. Pfuller, J. Braun, F. Paul, and J. Wuerfel. 2012. Brain viscoelasticity alteration in chronic-progressive multiple sclerosis. *PLoS One* 7 (1):e29888. doi:10.1371/journal.pone.0029888.

Streitberger, K. J., E. Wiener, J. Hoffmann, F. B. Freimann, D. Klatt, J. Braun, K. Lin et al. 2011. In vivo viscoelastic properties of the brain in normal pressure hydrocephalus. *NMR Biomed* 24 (4):385–392. doi:10.1002/nbm.1602.

Su, L. N., S. L. Guo, B. X. Li, and P. Yang. 2014. Diagnostic value of magnetic resonance elastography for detecting and staging of hepatic fibrosis: a meta-analysis. *Clin Radiol* 69 (12):e545–e552. doi:10.1016/j.crad.2014.09.001.

Talwalkar, J. A., M. Yin, S. Venkatesh, P. J. Rossman, R. C. Grimm, A. Manduca, A. Romano, P. S. Kamath, and R. L. Ehman. 2009. Feasibility of in vivo MR elastographic splenic stiffness measurements in the assessment of portal hypertension. *AJR Am J Roentgenol* 193 (1):122–127. doi:10.2214/ajr.07.3504.

Thormer, G., M. Reiss-Zimmermann, J. Otto, K. T. Hoffmann, M. Moche, N. Garnov, T. Kahn, and H. Busse. 2013. Novel technique for MR elastography of the prostate using a modified standard endorectal coil as actuator. *J Magn Reson Imaging* 37 (6):1480–1485. doi:10.1002/jmri.23850.

Trout, A. T., S. Serai, A. D. Mahley, H. Wang, Y. Zhang, B. Zhang, and J. R. Dillman. 2016. Liver stiffness measurements with MR elastography: Agreement and repeatability across imaging systems, field strengths, and pulse sequences. *Radiology* 281:793–804. doi:10.1148/radiol.2016160209.

Tse, Z. T., H. Janssen, A. Hamed, M. Ristic, I. Young, and M. Lamperth. 2009. Magnetic resonance elastography hardware design: A survey. *Proc Inst Mech Eng H* 223 (4):497–514.

Tzschatzsch, H., J. Guo, F. Dittmann, S. Hirsch, E. Barnhill, K. Johrens, J. Braun, and I. Sack. 2016. Tomoelastography by multifrequency wave number recovery from time-harmonic propagating shear waves. *Med Image Anal* 30:1–10. doi:10.1016/j.media.2016.01.001.

Van Houten, E. E., K. D. Paulsen, M. I. Miga, F. E. Kennedy, and J. B. Weaver. 1999. An overlapping subzone technique for MR-based elastic property reconstruction. *Magn Reson Med* 42 (4):779–786.

van Soest, G., F. Mastik, N. de Jong, and A. F. van der Steen. 2007. Robust intravascular optical coherence elastography by line correlations. *Phys Med Biol* 52 (9):2445–2458. doi:10.1088/0031-9155/52/9/008.

Van Houten, E. E. W., M. M. Doyley, F. E. Kennedy, J. B. Weaver, and K. D. Paulsen. 2003. Initial in vivo experience with steady-state subzone-based MR elastography of the human breast. *J Magn Reson Imaging* 17:72–85.

Venkatesh, S. K., and R. L. Ehman. 2015. Magnetic resonance elastography of abdomen. *Abdom Imaging* 40 (4):745–759. doi:10.1007/s00261-014-0315-6.

Venkatesh, S. K., M. Yin, and R. L. Ehman. 2013. Magnetic resonance elastography of liver: Technique, analysis, and clinical applications. *J Magn Reson Imaging* 37 (3):544–555. doi:10.1002/jmri.23731.

Venkatesh, S. K., M. Yin, J. F. Glockner, N. Takahashi, P. A. Araoz, J. A. Talwalkar, and R. L. Ehman. 2008. MR elastography of liver tumors: Preliminary results. *Amer J Roent* 190:1534–1540.

Wang, R. K., Z. Ma, and S. J. Kirkpatrick. 2006. Tissue Doppler optical coherence elastography for real time strain rate and strain mapping of soft tissue. *Appl Phys Lett* 89 (14):144103. doi:10.1063/1.2357854.

Weaver, J. B., A. J. Pattison, M. D. McGarry, I. M. Perreard, J. G. Swienckowski, C. J. Eskey, S. S. Lollis, and K. D. Paulsen. 2012. Brain mechanical property measurement using MRE with intrinsic activation. *Phys Med Biol* 57 (22):7275–7287. doi:10.1088/0031-9155/57/22/7275.

Wu, T., J. P. Felmlee, J. F. Greenleaf, S. J. Riederer, and R. L. Ehman. 2000. MR imaging of shear waves generated by focused ultrasound. *Magn Reson Med* 43 (1):111–115.

Wuerfel, J., F. Paul, B. Beierbach, U. Hamhaber, D. Klatt, S. Papazoglou, F. Zipp, P. Martus, J. Braun, and I. Sack. 2010. MR-elastography reveals degradation of tissue integrity in multiple sclerosis. *Neuroimage* 49 (3):2520–2525. doi:10.1016/j.neuroimage.2009.06.018.

Xu, H., S. F. Othman, L. Hong, I. A. Peptan, and R. L. Magin. 2006. Magnetic resonance microscopy for monitoring osteogenesis in tissue-engineered construct in vitro. *Phys Med Biol* 51 (3):719–732. doi:10.1088/0031-9155/51/3/016.

Xu, L., J. Chen, K. J. Glaser, M. Yin, P. J. Rossman, and R. L. Ehman. 2013. MR elastography of the human abdominal aorta: A preliminary study. *J Magn Reson Imaging* 38 (6):1549–1553. doi:10.1002/jmri.24056.

Yeung, D. K., K. S. Bhatia, Y. Y. Lee, A. D. King, P. Garteiser, R. Sinkus, and A. T. Ahuja. 2013. MR elastography of the head and neck: Driver design and initial results. *Magn Reson Imaging* 31 (4):624–629. doi:10.1016/j.mri.2012.09.008.

Yin, M., J. A. Talwalkar, K. J. Glaser, A. Manduca, R. C. Grimm, P. J. Rossman, J. L. Fidler, and R. L. Ehman. 2007a. Assessment of hepatic fibrosis with magnetic resonance elastography. *Clin Gastroenterol Hepatol* 5:1207–1213.

Yin, M., J. A. Talwalkar, K. J. Glaser, A. Manduca, R. C. Grimm, P. J. Rossman, J. L. Fidler, and R. L. Ehman. 2007b. Assessment of hepatic fibrosis with magnetic resonance elastography. *Clin Gastroenterol Hepatol* 5 (10):1207–1213.e2. doi:10.1016/j.cgh.2007.06.012.

Yin, M., J. Woollard, X. Wang, V. Torres, P. C. Harris, C. J. Ward, K. J. Glaser, A. Manduca, and R. L. Ehman. 2007c. Quantitative assessment of hepatic fibrosis in an animal model with magnetic resonance elastography. *Magn Reson Med* 58:346–353.

Yoshimitsu, K., Y. Shinagawa, T. Mitsufuji, E. Mutoh, H. Urakawa, K. Sakamoto, R. Fujimitsu, and K. Takano. 2016. Preliminary comparison of multi-scale and multi-model direct inversion algorithms for 3T MR elastography. *Magn Reson Med Sci* 16:73–77. doi:10.2463/mrms.mp.2016-0047.

Zerhouni, E. A., D. M. Parish, W. J. Rogers, A. Yang, and E. P. Shapiro. 1988. Human heart: tagging with MR imaging—A method for noninvasive assessment of myocardial motion. *Radiology* 169 (1):59–63. doi:10.1148/radiology.169.1.3420283.

4

Biomechanics of Cancer

Homeyra
Pourmohammadali,
Mohammad
Kohandel, and
Sivabal
Sivaloganathan

4.1 Introduction

The past decade has witnessed an increasing symbiotic interaction between the mathematical and biomedical sciences. In particular, this has given rise to the very fruitful interdisciplinary field of mathematical oncology—a field that, in recent years, has contributed (in no small part) to notable advances in clinical oncology and our understanding of carcinogenesis.

The human organism contains a wide variety of soft tissues in our body (tendons, ligaments, blood vessels, etc.), which have evolved over the course of millennia to provide support and protection to critical structures and organs of the human body. The soft tissues are highly flexible with complex composite structures reinforced by fibers. They are deformable, anisotropic, and mostly inhomogeneous with viscoelastic behavior and nonlinear tensile responses. Their soft mechanical properties are primarily due to the concentration and arrangement of the main components such as collagen, elastin, or proteoglycans (Holzapfel, 2000). Quantification of the mechanical properties, either at the tissue or cell level can be useful in cancer diagnosis, monitoring some therapies and in clinical decision-making.

Cancer cells experience a variety of mechanical forces from their neighboring cells and through the extracellular matrices, or as a result of externally applied mechanical loads such as compression, shear, tension, flow-induced shear, or interstitial stresses. Cancer cells and colonies also exert mechanical forces on their surroundings. These forces can be transformed into cellular responses through mechanotransduction and can alter the cell characteristics, including proliferation, morphogenesis, gene expression, and synthesis of proteins. The migration and differentiation of cells within living tissues are also affected by these biomechanical forces. Therefore, the substantial alteration in the biomechanical properties or the heterogeneity (the quality or state of being diverse in character, content, or elements) within tissues can act as indicators of cancer diagnosis and its progression. The functional role of some of the typical biomechanical forces, such as compression, shear, interstitial fluid flow, and internal and external loads in cancer are briefly discussed in the following sections.

4.1.1 Compression

As cancer cells grow uncontrollably in a limited space, compressive forces are generated inside the tumor. The increase in growth-induced stress results in significant deformation of the tumor microenvironment. Although the intratumoral region is in compression, the periphery of a tumor is in tension due to shear stresses. The peripheral tensile stress provides a smooth transition between high-compressive stresses inside the tumor and stresses within the surrounding normal tissue (Jain et al., 2014). In some cases, the resulting intratumoral compression can be escalated to a level that leads to pinching and occlusion of blood and lymphatic vessels. The compression of blood and lymphatic vessels limits the delivery of nutrient, oxygen, and drugs and can create a hypoxic microenvironment (Padera et al., 2004). The lack of nutrient and oxygen supply may also cause the formation of necrotic tissues inside the tumor. At the same time, this compression is believed to protect cancer cells from the immune system of the host, allowing them to access oxygen and nutrients that are essential for tumor progression (Jain, 2013). The compression of blood vessels causes the blood flow to be decreased and vascular shunts to be formed and limits drug delivery to certain regions (especially in large tumors).

The compression also has an effect on the extracellular matrix (ECM), which, in turn, exerts forces that are capable of changing cell shape and cytoskeletal organization. This influences the proliferation and apoptosis of tumor cells (Cheng et al., 2009), remodels the cytoskeleton organization, and has an impact on the migration process of cancer cells (Tse et al., 2012). The compressed cells undergo a phenotypic transformation, become leader cells, adhere better to the substrate (due to larger cell–substrate contact areas), and participate in a collective or coordinated migration (Friedl and Gilmour, 2009). The presence of a moderate level of continuous compressive forces is beneficial for cell motility and enhanced migration; however, excessive amounts are detrimental due to the resultant increased cell death, which is an impediment to cell migration (Tse et al., 2012). Cell compression changes the stromal cell function, the synthesis and organization of ECM, the gene expression (Demou, 2010), and the invasiveness of the cells (Tse et al., 2012; Jain et al., 2014).

4.1.2 Shear Stress and Interstitial Fluid Flow/Pressure

The circulating tumor cells (CTC) are the main cause of distal metastasis, and the indications are that mechanotransduction may play a crucial role in the epithelial–mesenchymal transition (EMT), which has been demonstrated experimentally to increase the metastatic potential of cancer cells significantly. Cancer cells are exposed to a hemodynamic microenvironment with shear stresses generated by interstitial flows (Mitchell and King, 2013). The interstitial flow velocity that is in the range of 0.1–1.0 µm/s in normal tissues can increase to about 4 µm/s in cancerous tissue (Swartz and Lund, 2012). The fluid shear stress is estimated to be 0.007–0.015 dyn/cm² for interstitial flow of 1 µm/s (Pedersen et al., 2007) and is significantly more during circulation, where it increases to 0.5–4.0 dyn/cm² in the venous circulation and to 4.0–30.0 dyn/cm² in the arterial circulation (Mitchell and King, 2013). Therefore, the response of tumor cells to a wide range of shear stresses is important and plays an important role in cancer metastasis. Despite the fact that higher levels of fluid shear stress decrease the survival of CTCs, at the same time, these stresses help in the dynamic rolling of these cells and let them bind more easily to the vascular endothelium (Burdick et al. 2003).

Studies of the effects of fluid shear stress on the proliferation and viability of CTCs indicate that the hemodynamic shear stress applied in a certain range can destroy CTCs in the bloodstream. Human arteries on an average experience shear stresses of 15 dyn/cm² and veins at a resting state experience 1–6 dyn/cm². This can increase during exercise to higher level of 60 dyn/cm² in the femoral arteries. The flow-induced shear stress modulates cell-cycle distribution. Tumor cell lines under shear stress (~12 dyn/cm² for 24–48 h) induce G2/M cell-cycle arrest, whereas those under static conditions go through G0/G1 arrest (Chang et al., 2008).

4.1.3 External and Internal Forces

When a growing tumor deforms the surrounding normal tissue, it generates an external stress, in addition to the normal and shear stresses within the tumor. This stress either directly compresses cancer and stromal cells and ECM constituents, or indirectly deforms blood and lymphatic vessels. The external stresses can be utilized as an indicator of tumor morphology. The loads may be isotropic; in which case, this would stimulate spherical tumor growth. On the other hand, anisotropic loads would lead to asymmetric tumor growth with a tendency for cancer cells to grow in the direction that minimizes stress (Jain et al., 2014). The alteration of the stress field can affect the aggregate remodeling and tumor spheroid growth pattern. For instance, the growth of cancer cells in agarose of varying concentration can lead to varying external stresses on the cells. The higher the agarose concentration, the greater the external stress on the cells.

Depending on the concentration, the external stress on the spheroids can change between 28–120 mmHg (3.7–16.0 kPa) (Cheng et al., 2009). Anisotropic mechanical loads generally result in apoptosis in high-compressive stress regions and proliferation in low-stress regions (Jain et al., 2014).

4.2 The Effects of Biomechanical Forces on Cancer Cells

The effects of biomechanical forces on cancer cells are discussed in the following sections covering different aspects, including cell deformation, adhesion, migration, colonization, heterogeneity, phenotype change, apoptosis, tumor growth, and metastasis.

4.2.1 Cell Deformation

The biomechanical properties of cells change as they become cancerous or are surrounded by malignant tumor cells. The amount of cell deformation or cell stiffness can be an indicator to distinguish healthy and diseased cells (Hou et al., 2009; Xu et al., 2012). The cytoskeletal structure and concentration of cytoskeletal content alters during cancer. The rigid and ordered structure of the cytoskeleton becomes compliant and irregular due to forces generated by changes in the mechanical properties of the cells (Suresh, 2007). The deformability or compliance of tumor cells often increases, and their stiffness and elasticity decrease in the cancerous state (Katira et al., 2013). The higher deformability correlates with higher malignancy and ability of the tumor cells to metastasize to distant regions of the body. Unlike cells that become softer, the ECM becomes stiffer when cancer cells become more aggressive. However, despite the dissimilarity of the stiffness of cancer cells and their surroundings, they have similar effects on the proliferation rate of the cells according to results from computational modeling (Klein et al., 2009). The model correlates the mechanical regulation of cells with the cell shapes and sets a threshold for the size of cell clusters. The lower the cell stiffness, the more this gives rise to uncontrolled growth of the cells (Katira et al., 2012). Metastatic potential of the soft tumor cells may be due to facilitation of the migration process (due to less cell stiffness), especially through narrower arteries and sharper turns (Drasdo and Hoehme, 2012), which is also facilitated through the EMT process.

4.2.2 Cell Adhesion

The ability of cells to bind with other normal and cancerous cells or to the surrounding ECM changes during carcinogenesis. This ability can change depending on the stage of cancer progression and on the cell phenotype (Katira et al., 2013). One of the important transmembrane proteins that has a functional role in cell adhesion is cadherin. The increasing malignancy results in more P-cadherin binding between cells (Van Marck et al., 2005) and less E-cadherin-mediated adhesion (Bryan et al., 2008). Mathematical models have been used to study the effect of cell-adhesion changes, in particular the relationship between internal pressure of the tumor and surface tension, which is a function of cell–cell adhesion (Byrne and Chaplain, 1996).

4.2.3 Cell Migration

Migration through the connective tissues is a prerequisite for metastasis (Mierke et al., 2008). Cell migration or invasion is mostly a mechanical process, which includes changes in cell adhesion and cell shape, cell movement and force generation. During the migration process, the motile cell becomes elongated in morphology first and then by means of extension of the cell's leading edge, it attaches to the ECM substrate. The attachment causes the entire cell body or some region of the leading edge to contract and create a traction force. The resulting force causes the cell body and its trailing edge to glide forward (Friedl and Wolf, 2003).

The force-generation mechanism throughout the migration process is still not fully understood. Some of the main questions relate to how the cells push or pull, how they initiate and sustain motion (how they adhere to the tissue matrix), or what magnitude of adhesive forces are necessary to maintain this process (Mierke et al., 2008). Experimental measurement of cell-traction forces has been one of the main approaches used to try and answer some of these questions. In 2D experiments with traction microscopy (Sabass et al., 2008), the adherent cells are plated on an elastic substrate with known elastic modulus. The cells that adhere to the substrate try to spread and generate traction forces that deform the substrate. Then, using continuum mechanics theory, together with the measured substrate deformation, the traction forces are computed. The current thinking suggests that dynamic regulation of adhesion receptor clustering, the formation of focal adhesion, and remodeling of cytoskeletal architecture cause the cells to sense the ECM stiffness and correspondingly to respond to it. Mounting experimental evidence indicates that the biomechanical properties of the ECM affect the contractile force and migration of the cells (Mierke et al., 2008). The migration speed is highly dependent on the adhesion and de-adhesion turnover rate, or in other words, the focal contact strength and migration speed are inversely proportional. The more that the focal contact strength stabilizes, the less the tumor cells detach and the lower the migration rate is. Despite the usefulness of 2D experiments in explaining some of the cellular behavior in the migration process, the dynamic environment of tumor cells and therefore the traction forces can substantially change in a 3D culture (Zaman et al., 2006).

4.2.4 Colonization

Colonization represents the ability of a single tumor cell to grow into a colony or to undergo unlimited division. Clonogenic assays, which measure the fractions of cells that are capable of colony formation, are mostly to determine the cell reproductive and survival fraction after treatment or the effectiveness of cytotoxic agents. Some of these assays were performed under additional mechanical loading conditions. For instance, the experiment on esophageal cancer cells under various flow conditions was done using channels of microfluidic chips (Calibasi Kocal et al., 2016). The seeded cells in each channel were either under static control or fluid flow conditions during incubation.

The study used aldehyde dehydrogenase (ALDH) activity as a marker of colony-forming ability or tumor metastasis. Generally, a significant increase in ALDH activity indicates more prominent stem or progenitor properties, a greater capacity for self-renewal and tumorigenesis, and greater heterogeneity recapitulating the parental tumor (Rodriguez-torres and Allan, 2016). Under static conditions, the ALDH-positive cells had about 4.4%–10.8% increase in their population, whereas under flow conditions, they had 6.0%–20.9% increase (Calibasi Kocal et al., 2016). This translates to a 2.2-fold higher capacity for colony formation under flow conditions than under static controls. Despite the fact that primary tumors have a tendency to metastasize to distal sites, autopsy data analysis shows that metastatic sites are not colonized randomly and that there are certain preferential sites (Weiss, 1992; Ginestier et al., 2007).

4.2.5 Heterogeneity

A tumor is a complex heterogeneous structure containing blood vessels, ECM, and different cell types and cell populations. There are two main proposed models to explain heterogeneity within cancer cells (Prasetyanti and Medema, 2017): one is the so-called clonal evolution model that explains the heterogeneity as arising through a process of natural selection, where stochastic mutations in individual cells give rise to the emergence of a *fittest* clone (Anderson et al., 2006; Sottoriva et al., 2010; Waclaw et al., 2015). The other is the cancer stem cell (CSC) hypothesis that posits the existence of a differentiation hierarchy that has its origins in healthy stem cells. On the basis of the CSC hypothesis, tumor progression is primarily due to the activity of only a small proportion of tumor cells known as CSCs. These stem cells have the capacity for self-renewal and differentiation into nonstem cells that populate the bulk of the tumor. From this perspective, tumor cells are the result of differentiation of CSCs (Clevers, 2011). The clonal and CSC concepts are not mutually exclusive, and tumor growth, in reality, may be driven by a combination of both of these proposed mechanisms, in addition to others.

Intratumor heterogeneity can arise as a result of many microenvironmental factors, including biological and chemical regulators, in addition to physical factors such as biomechanical forces (Prasetyanti and Medema, 2017). The level of heterogeneity and plasticity of cancer cells can vary depending on these biomechanical conditions. The selective pressures applied to tumor cells during cancer therapy can also change intratumor heterogeneity. In the case of breast cancer, for instance, intratumor heterogeneity may be reduced as a result of therapy that tends to act as a selective pressure (Marusyk et al., 2014; Calibasi Kocal et al., 2016).

4.2.6 Tumor Growth

The biomechanical forces generated due to uncontrolled expansion of tumor cells in limited space have a profound effect on cellular behavior, stimulate the migration of carcinoma cells, and influence both tumor growth and progression (Tse et al., 2012). The role of various mechanical signals on cell behavior has been highlighted in some studies. These studies showed that an increase in the interstitial fluid pressure (Boucher and Jain, 1992), interstitial fluid flow (Ng et al., 2005), compression, or elevation of ECM stiffness (Shao et al., 2000) has a pronounced impact on the development and growth of tumors. Furthermore, the microenvironment, which includes the surrounding cells, ECM, and soluble factors in the extracellular environment, plays a critical role in tumor development. The growth and progression of a tumor also contribute to further physical changes due to the creation of abnormal forces and fluid flows. Many of the changes that occur simultaneously in the material and structural properties of the tumor microenvironment serve to amplify the growth-induced stresses as well.

The uncontrolled expansion of cells increases the stress within a tumor and the radial and circumferential stresses in the surrounding tissues. The stress and strain can exceed 10 kPa and 40%, respectively, demonstrated experimentally in *in vitro* growth of tumor spheroids in agarose (Helmlinger et al., 1997; Shieh, 2011; Jain et al., 2014). Several studies have reported correlation of stiffer tissues with growth of tumors in some type of cancers such as breast, liver, pancreatic, and prostate cancers (Hoyt et al., 2008; Boyd et al., 2009; Gang et al., 2009; Nahon et al., 2009; Iglesias–Garcia et al., 2010; Zhao et al., 2010). The higher stiffness is because of larger amounts of stromal tissue with stiffer ECM than normal (Shieh, 2011).

Tumor growth is associated with significant changes and fluctuations in the tumor microenvironment. After tumor cells expand their numbers in an uncontrolled fashion and the tumor size exceeds a certain size, on the order of a few millimeters, the solid tumor becomes hypoxic and starts releasing tumor angiogenic factors (TAFs) to stimulate an angiogenic response from the host tissue (Boucher et al., 1996). This results in a highly irregular and abnormal network of vessels with higher permeability and leakiness than a normal, mature microvasculature network (Fukumura et al., 2010). The leakiness leads to a buildup of fluid and macromolecules within the tumor interstitium and gives rise to an increase in the tumor interstitial fluid pressure (Boucher and Jain, 1992; Harrell et al., 2007; Ruddell et al., 2008).

4.2.7 Metastasis and Progression

Metastasis is the term coined to describe the process by which tumor cells from a primary tumor in one location of the host organism colonize and give rise to a secondary tumor at another often distant site. An important step in metastasis is the EMT, which involves a transition of the colonizing cancer cells from an epithelial to mesenchymal phenotype, which renders the cells more motile. Furthermore, experimental studies show that cancer cells that undergo this transition display many of the characteristics of CSCs (Mani et al., 2008). After detachment from the primary tumor, cancer cells start invading other tissues by first breaking through the underlying basement membrane, passing through the interstitial connective tissues, and then enter into the blood vessel and blood circulation through the vascular basement membrane through the process known as intravasation. After intravasation, some of the CTC adhere to the wall of the blood vessel and extravasate through the wall, migrate into the normal local tissue, and form another tumor. Each of these steps are highly dependent on the physical interactions and mechanical forces between cancer cells and the microenvironment (Wirtz et al., 2011). A favorable local microenvironment spurs the process of metastasis but from a mechanical point of view it is believed that it happens at sites of optimal flow patterns, suggesting the functional role of mechanical cues in the initial delivery and arrest of tumor cells (Azevedo et al., 2015).

Intravasation and extravasation require cancer cells to highly deform to get into and out of the endothelial cell layer of the vessel wall. The significant change in the shape (phenotype) is driven by cytoskeletal remodeling that facilitates the entry of tumor cells into the endothelial cell–cell junctions surrounding blood vessels. At such high-deformation rate, the cytoplasm has an elastic behavior and acts more similar to a viscous material. The more challenging part of the migration process of tumor cells may be the deformation of the interphase endothelium (the largest organelle in the cell that is 10 times stiffer than the cytoplasm) (Tseng et al., 2004). Therefore, optimal mechanical properties are required in order for the tumor cells to migrate. The cells are unable to deform the cross-linked collagen fibers of the matrix and thus unable to effectively migrate if they are too soft or too stiff. Even a single cell of a specific cell type can behave in a heterogeneous fashion and present a wide range of mechanical properties at different steps of the metastatic cascade. The dynamic changes in these properties allow the cells to survive in the harsh and changing environment of the stromal space and blood and lymphatic vessels. Other factors such as interstitial flows (Swartz and Fleury, 2007), electric fields (Mysielka and Djamgoz, 2004), and biochemical gradients may also impact and influence changes in these mechanical properties at various stages.

Cancer cells are softer (with lower stiffness) than normal cells and this compliance correlates with an increased metastatic potential (Cross et al., 2007). The CTC, as the main source of metastasis initiating cells (Azevedo et al., 2015), also have to strike a subtle balance between the cell velocity and adhesion in order to bind to the vessel wall and initiate the process of extravasation. As they transit into the circulation system, they collide with host cells (blood and endothelial cells) and are exposed to hemodynamic forces. The CTC that seeds a secondary tumor has to successfully navigate through these shear forces, adhere to the vessel wall at the secondary site, and exit the vasculature in order to be able to initiate a secondary tumor. Better understanding of the mechanism of cancer progression and the function of physical interactions and mechanical forces in metastasis can lay the foundation to guide newer therapeutic approaches.

4.3 Theoretical Models Used for Biomechanical Studies of Cancer

The biomechanical response of solid tumors can be studied and simulated (at multiple scales ranging from the molecular scale, all the way through to the macroscopic tissue-level scale) using mathematical/computational models. An understanding of the interaction and interrelationship between many cancer-related processes has been facilitated by these models (which vary widely from continuum-based to discrete agent-based models (Galle et al., 2006; Sanga et al., 2007; Byrne and Drasdo, 2009; Rejniak and McCawley, 2010; Deisboeck et al., 2011; Frieboes et al., 2011; Kam et al., 2012). In the continuum-based model, tumor mass density, interstitial fluid pressure, interstitial filtration velocity, and other quantities associated with the tumor microenvironment are considered to be scalar or vector fields evolving over 3D domains. In the discrete-model approach, the system is represented by individual cells that interact with each other and with their surroundings. Hybrid models have also been proposed as multiscale models, which attempt to combine the two continuum and discrete approaches. In the hybrid model, the continuum approach is used to capture variations in macroscopic quantities (tumor mass, interstitial pressure, etc.) and contributions from the cellular scale are linked in a phenomenological manner with a discrete model that purportedly captures cellular level interactions. Further details of these approaches can be found in the following research articles: (1) Galle et al., 2006; (2) Sanga et al., 2007; (3) Byrne and Drasdo, 2009; (4) Stolarska et al., 2009; (5) Rejniak and McCawley, 2010; (6) Deisboeck et al., 2011; and (7) Frieboes et al., 2011; Kam et al., 2012, among others. Various aspects of cancer have been successfully investigated using these approaches. These include papers focusing on carcinogenesis and cancer progression, emergence of heterogeneity, phenotypic alterations, nutrient/chemical gradients, external and internal forces in tumors, and mechanical interaction between cells, ECM, and their surroundings. We have addressed many of these aspects in very broad terms, while those interested in particular aspects will find further details of the mathematical and mechanical models in the detailed references provided. The focus of our discussion, however, is on models used to calculate the magnitude of various mechanical/biomechanical stresses in tumors (Voutouri et al., 2014), and to estimate their mechanical responses (Ambrosi and Preziosi, 2002; Byrne and Preziosi, 2003; Roose et al., 2003; Sarntinoranont et al., 2003).

Some of the early mathematical models for tumor evolution (Greenspan, 1976; McElwain and Morris, 1978; Adam, 1987) were mainly temporal (ODE models) for only one population of tumor cells. These were developed with assumptions such as unidirectionality of tumor growth and constant tumor density. As with all model development, it became clear that more sophisticated spatiotemporal models were required to more accurately capture mechanisms driving carcinogenesis. The foundation of many of these types of models has been based on physical conservation laws: (1) mass-balance equations for cells, and (2) conservation of momentum (fluid flow equation), and (3) energy conservation. In many instances, these give rise to reaction-diffusion type equations, but other approaches have utilized the continuum mechanics-based theory of poroelasticity and have successfully used this to explain the rise in interstitial pressure in tumors during tumorigenesis (Baxter and Jain, 1989; Phipps and Kohandel, 2011; Burazin et al., 2017). The assumptions based on experimental observations are often made to obtain a closed system of equations. The resulting models often include reaction-diffusion equations and sometimes integro-differential equations. Reaction-diffusion equations are typically used to model the distribution of nutrients and chemicals such as oxygen, glucose, or drugs within a tumor and integro-differential equations usually arise in trying to capture the rate of volumetric growth of a tumor. Despite success in application of these models to a few *in vitro* studies (Sutherland, 1988; Kunz-Schughart, 2000) by and large they have failed to describe 3D tumor developments with more than one population of cell types primarily due to the challenge of closing the mass-balance equation system. Further assumptions such as similar velocities for motile cells and assumptions of radial symmetry were often made to close the system (Ward et al., 1997; Ward and King, 1999).

The failure of many of these models to fully capture the effects of mechanical forces on the tumor growth has provided the impetus to develop different approaches. One direction has been in the development of multiphase models (Drew and Segel, 1971; Fowler, 1997) in which mass and moment-balance equations for each population of cells are derived, taking into consideration cell–cell mechanical interactions and applying constitutive laws. The use of constitutive laws in the theory of mixtures for a two-phase model enabled the calculations of intratumoral stresses by using homogenization theory to combine in a self-consistent manner the contributions of the solid and liquid phases. The challenge of closing the equations of the mass and momentum balance was later resolved (Ambrosi and Preziosi, 2002) by relating the pressure of different cell populations to their respective velocities using Darcy's law. Therefore, the inclusion of mass exchange between two phases (solid and liquid phases), and the dependency of the cell-proliferation rate on cellular stress were two crucial mechanistic assumptions that were incorporated in the two-phase model. The sensitivity of the equilibrium size of a tumor to variations in parameters of the model (such as maximum cell-proliferation rate and the rate of natural cell death) was explored using both numerical and analytical methods. The model helped to explain, to some degree, how externally applied stresses can affect the tumor development and how the dependency of cell proliferation on cellular stress impacts the results. The model predicted the suppressive effect of mechanical stresses in decreasing the equilibrium size of a tumor and increasing mitosis.

Continuum mechanics provides a framework to quantify and model the phenomenological responses of biological systems. With some modifications, the mechanics of cell deformation, subcellular components, molecular networks, and attachment system can be modeled using a continuum-mechanics approach. Living cells can be modeled as a continuum that either contains an elastic cortex that surrounds the viscoelastic fluid (Schmid-Schonbein et al., 1995) or includes an elastic nucleus in a viscous cytoplasm (Dong et al., 1991). The elements that experience loads in the continuum models are significantly smaller than the cell size that limits our insight into the contribution of distinct molecular structures to cell mechanics. Thus, cell-function regulation by mechanical forces remains unclear in the absence of a clear link between mechanics, microstructure, and molecular biology (Wang et al., 2001).

Tensegrity (tensional integrity), which is a structural principle proposed by Fuller in 1960s (Fuller, 1961, 1964), has also been used for modeling the structures of living tissues. Using this principle, continuous tension and discontinuous compression participate in stabilizing structures or the tensegrity of the systems (Tadeo et al., 2014). Tensegrity structures are self-stabilized and enable the transfer of applied loads throughout their structures while minimizing the damage to it. Tensegrity concepts have been applied to complex biological organisms as well. The application of this principle to the tensegrity structures of living bodies (such as the musculoskeletal system, proteins, or DNA) has given rise to the new field of biotensegrity (Ingber, 1994; Randel, 2013; Swanson, 2013; Tadeo et al., 2014). As biological tissues display viscoelastic properties, this has given rise to models that use the mathematical theory of viscoelasticity to capture changes in biological tissues. The simple standard linear viscoelastic model has been used to quantify the mechanical properties of cells and cytoskeleton networks using combination of dashpot and spring with a wide range of frequencies. It has also been used to capture the creep response of different cells such as leukocytes, neutrophils, and smooth muscle cells (Drury and Dembo, 2001; Liao et al., 2006). Some viscoelastic models, reviewed in Suresh (2007) for quantifying the mechanical properties of biological tissues include the following: (a) classical Voigt and Maxwell models to study cell and cytoskeletal deformations by quantifying the viscoelastic response of soft materials and creeping solids, (b) classical Hertzian elastic indentation model (Radmacher et al., 1996), and (c) elasticity model (Bao and Suresh, 2003).

4.4 Experimental Assays for Biomechanical Forces of Cancer Cells

The biomechanics and mechanobiology of living cells can be investigated using *in vitro* experimental techniques (Addae-Mensah and Wikswo, 2008). The particular technique used will depend on the mechanical property of interest and also on the range of accessible measurements. In a physiological environment, cells experience a broad spectrum of forces, yet technological advances have provided some opportunity for analysis of their mechanical properties even at the very small nano- to pico-scales. The experimental techniques are classified differently and are categorized based on whether the technique is active or passive (Addae-Mensah and Wikswo, 2008). In the case of active techniques, various forces are applied to deform cells in a particular manner; in contrast, for passive methods, only the mechanical forces such as traction forces generated by the cells themselves are sensed. Some of the current active methods include the following: (a) atomic force microscopy (AFM) (Radmacher et al., 1996; Hessler et al., 2005; Lee and Lim, 2007), (b) micropipette aspiration, (c) optical tweezers (Mills et al., 2004; Li et al., 2005), (d) micromachined fore sensors and actuators, (e) shear flow methods, (f) stretching devices, and (g) carbon fiber-based systems. Current passive methods include the following: (a) elastic substratum method, (b) flexible sheets with embedded beads, (c) flexible sheets with micropipetted dots or grids, (d) micromachined cantilever beam, and (e) array of vertical microcantilevers. Some of these techniques such as optical tweezers, microplate stretcher (Thoumine and Ott, 1997; Thoumine et al., 1999), and micropipette aspiration are used for the measurement of single-cell deformation, whereas techniques such as AFM, magnetic twisting cytometry are used for measurement of forces in subcellular regions.

It is still far from clear which of these above-mentioned experimental methods for characterization of the biomechanical properties of normal living cells are most appropriate for the characterization of cancer cells and how the phenotype of the cancer cell or cancer cell line might influence the choice of measurement techniques. However, it is evident that the overwhelming majority of these techniques are capable of measuring only one or two parameters, such as the deformation or surface friction. The fact that tumors comprise a heterogeneous population of cells clearly necessitates the use or development of techniques that can simultaneously measure multiple parameters.

4.5 The Application of Biomechanical Information in Cancer Therapies

The selection of appropriate therapeutic strategies in the treatment of cancer is primarily determined by the type and stage of the cancer. Surgery, radiation therapy, and chemotherapy represent the standard classical treatment modalities; however, in recent years, immunotherapy, targeted therapy, hormone therapy, stem cell transplant, and precision medicine have emerged as novel powerful therapeutic strategies (Emens et al., 2017; Hard et al., 2017). Tumors can be partially or fully resected from a patient's body through surgery; the size and bulk of tumors are often reduced through radiation therapy or chemotherapy prior to surgery (neoadjuvant radio/chemotherapies), or any remaining cancer cells are eradicated through radiation/chemotherapy postsurgery (adjuvant radio/chemotherapies). Of the newer approaches, immunotherapy holds much promise, because it primes the immune system of patients themselves to attack and eradicate the cancer. The alterations in the growth and differentiation of cancer cells can be monitored in this type of targeted therapy. Some types of cancers, such as breast and prostate cancers can be controlled by hormone therapy. Another approach for patients who have lost hematopoietic stem cells through radiation and chemotherapies is to have these replenished through stem cell transplant. If the genetic basis of a cancer is known, precision medicine can be used to deliver targeted treatment. Mechanotherapy is a novel optional treatment for tumors that are smaller and more on the surface tissue (Lee and Lim, 2007). Indeed, combination therapies of many of the above-mentioned therapies are often administered depending on the type of cancer and its stage and aggressivity.

4.5.1 The Role of Biomechanical Forces in Cancer Treatments

The structural and functional changes that give rise to many types of disease states can be unraveled through the skillful analysis of biomechanical data. This is particularly important in understanding the pathophysiology, genesis, and progression of cancers, where early diagnosis and detection can profoundly affect their prevention and control (Lee and Lim, 2007). Biomechanical data can be used for improving assays, for designing and increasing the detection accuracy of diagnostic devices, and other techniques for early cancer detection in the early stages when signs and symptoms are not fully physically manifested. Therefore, the mechanical characterization of cancer cells using direct or indirect measurements can be optimally used to modify and improve cancer therapies. In the direct approach, the mechanical characteristics such as elasticity and viscoelasticity of single cells are measured (Sun et al., 2003; Papi et al., 2009), whereas in the indirect approach, the mechanical characteristics such as cell deformability are measured (Sakuma et al., 2014; Tsai et al., 2014). Cell deformability is a cellular mechanical property that (in some sense) characterizes the carcinogenic state and is a potential biological marker candidate in the detection and diagnosis of cancer using mechanical probes.

4.5.2 Mechanotherapy for Cancer Treatment

Mechanotherapy is an emergent field that focuses on the therapeutic role and effect of mechanical forces. It is based on the premise that cells throughout their whole life are exposed to mechanical forces, which, in turn, must trigger signaling pathways that affect molecular-level activities, through mechanotransduction. Mechanotherapy can be used as a tool for cancer treatment by triggering cellular-level and molecular-level pathways in tumors through the judicious application of mechanical forces. In this framework, strategies are proposed with a view to changing the tumor microenvironment and enhancing the delivery of therapeutic agents to make the treatment process more effective (Jain et al., 2014). The fact that the mechanical environment of a tumor is highly affected by the stresses and by the presence of abnormal vessels lays the foundation for two strategies for improving drug delivery to tumors: (1) stress-alleviation and (2) vascular normalization (Stylianopoulos et al., 2012). The stress-alleviation strategy provides better perfusion and drug delivery by reducing the stress, increasing the diameter of intratumor lymphatic vessels, and reopening the compressed blood vessels (Chauhan et al., 2013). However, the vascular normalization strategy adjusts the blood flow rates in tumors and enhances perfusion by reducing the leakiness in tumors through changing the phenotype of the abnormal tumor vasculature to the functional phenotype of normal vessels (Jain, 2005; Goel et al., 2011). When tumor perfusion is increased, interstitial fluid pressure is decreased resulting in a greater driving pressure gradient across the vessel wall. This greatly facilitates the extravasation of drugs through the vessel wall. In addition, the increase in the size and number of openings in the vessel wall (Chauhan et al., 2012) facilitates the passage of more and larger particles and therapeutic agents through the vessel wall.

4.5.3 The Effects of Chemotherapy Drugs on Cell Mechanics

It is now becoming increasingly clear that the mechanical properties of cells are most likely changed by chemotherapeutic drugs during the process of cancer therapy. Many of these drugs alter the mechanical properties of cells, and hence their mechanical response and the state of the disease by (at a fundamental level) altering the cytoskeletal architecture of the cells (Grzanka et al., 2003). The study of the effects of chemotherapy on the mechanical properties of leukemia cells indicated a twofold increase in cell stiffness (Rosenbluth et al., 2006; Lam et al., 2007). This study was done using lymphoblastic and acute myeloid leukemia cells that were exposed to dexamethasone and daunorubicin drugs and their stiffness was measured using AFM. It was shown that the type of chemotherapy used affected the stiffness and the kinetics of cell death. Within an hour of exposure to chemotherapeutic drugs, the cell stiffness started to increase, and was significantly enhanced as a cell entered the process of apoptosis.

Studies carried out on breast cancer cells (MCF-7) exposed to the chemotherapeutic drug paclitaxel for 24 h, showed a significant increase in the cell elasticity, where elastic and viscous moduli both increased (Kaffas et al., 2013). The chemotherapy drug cisplatin affected the nanomechanics of ovarian cancer cells by increasing the stiffness of sensitive cells and disrupting the F-actin polymerization. Despite having no effect on the resistant cells, the findings of this study suggested that cancer-drug sensitivity can be revealed by nanomechanical profiling (Sharma et al., 2012). The Young's modulus decline (~60%) also occurred after 5 min of exposure to CB1 chemotherapy treatment for lung cancer cells (Kao et al., 2012).

4.6 Conclusion

There is perhaps no single area of human endeavor that has not been profoundly impacted by mathematics. In many senses, the biomedical sciences represent a *final frontier* where the full impact and benefits of the synergistic interaction with mathematics have yet to be realized. However, the nascent interdisciplinary field of mathematical oncology has been progressing in leaps and bounds over the past decade. Interdisciplinary teams comprised of mathematicians, clinical oncologists, and cancer biologists are now almost common place, and this has led to rapid and dramatic advances in both our understanding of the biology of cancer, and also in clinical oncology. We are optimistic that the coming years will see even greater advances and that the best is yet to come.

Acknowledgments

This work was financially supported by the Natural Sciences and Engineering Research Council of Canada (NSERC, discovery grant to SS and MK).

References

Adam, J. A. 1987. A mathematical model of tumor growth. II. Effects of geometry and spatial nonuniformity on stability. *Mathematical Biosciences*, 86(2), 183–211.

Addae-Mensah, K., and Wikswo, J. P. 2008. Measurement techniques for cellular biomechanics in vitro. *Experimental Biology and Medicine* (*Maywood*), 233(7), 792–809. doi:10.3181/0710-MR-278.

Ambrosi, D., and Preziosi, L. 2002. On the closure of mass balance models of tumour growth. *Mathematical Models and Methods in Applied Sciences*, 12, 737–754.

Anderson, A. R. A., Weaver, A. M., Cummings, P. T., and Quaranta, V. 2006. Tumor morphology and phenotypic evolution driven by selective pressure from the microenvironment. *Cell*, 127(5), 905–915.

Azevedo, A. S., Follain, G., Patthabhiraman, S., Harlem, S., and Goetz, J. G. 2015. Metastasis of circulating tumor cells: Favorable soil or suitable biomechanics, or both? *Cell Adhesion & Migration*, 9(5), 345–356.

Bao, G., and Suresh, S. 2003. Cell and molecular mechanics of biological materials. *Nature Materials*, 2(11), 715–725.

Baxter, L. T., and Jain, R. K. 1989. Transport of fluid and macromolecules in tumors. I. Role of interstitial pressure and convection. *Microvascular Research*, 37(1), 77–104. doi:10.1016/0026-2862(89)90074-5.

Boucher, Y., and Jain, R. K. 1992. Microvascular pressure is the principal driving force for interstitial hypertension in solid tumors: Implications for vascular collapse. *Cancer Research*, 52(18), 5110.

Boucher, Y., Leunig, M., and Jain, R. K. 1996. Tumor angiogenesis and interstitial hypertension. *Cancer Research*, 56(18), 4264–4266.

Boyd, N. F., Martin, L. J., Rommens, J. M., Paterson, A. D., Minkin, S., Yaffe, M. J., Stone, J., and Hopper, J. L. 2009. Mammographic density: A heritable risk factor for breast cancer. *Cancer Epidemiology: Modifiable Factors*, 343–360. doi:10.1007/978-1-60327-492-015.

Bryan, R. T., Atherfold, P. A., Yeo, Y. et al. 2008. Cadherin switching dictates the biology of transitional cell carcinoma of the bladder: Ex vivo and in vitro studies. *The Journal of Pathology*, 215(2), 184–194. doi:10.1002/path.2346.

Burazin, A., Tenti, G., Drapaca, C. S., and Sivaloganathan, S. Under review 2017. A poroelasticity theory approach to study the mechanisms leading to elevated interstitial pressure in solid tumours. *Bulletin of Mathematical Biology*,

Burdick, M. M., McCaffery, J. M., Kim, Y. S., Bochner, B. S., and Konstantopoulos, K. 2003. Colon carcinoma cell glycolipids, integrins, and other glycoproteins mediate adhesion to HUVECs under flow. *American Journal of Physiology. Cell Physiology*, 284(4), C977.

Byrne, H. M., and Chaplain, M. A. J. 1996. Modelling the role of cell-cell adhesion in the growth and development of carcinomas. *Mathematical and Computer Modelling*, 24(12), 1–17.

Byrne, H., and Drasdo, D. 2009. Individual-based and continuum models of growing cell populations: A comparison. *Journal of Mathematical Biology*, 58(4–5), 657–687.

Byrne, H., and Preziosi, L. 2003. Modelling solid tumour growth using the theory of mixtures. *Mathematical Medicine and Biology*, 20(4), 341–366.

Chang, S., Chang, C. A., Lee, D. et al. 2008. Tumor cell cycle arrest induced by shear stress: Roles of integrins and smad. *Proceedings of the National Academy of Sciences of the United States of America*, 105(10), 3927.

Chauhan, V. P., Martin, J. D., Liu, H. et al. 2013. Angiotensin inhibition enhances drug delivery and potentiates chemotherapy by decompressing tumour blood vessels. *Nature*, 4, 2516.

Chauhan, V. P., Stylianopoulos, T., Martin, J. D. et al. 2012. Normalization of tumour blood vessels improves the delivery of nanomedicines in a size-dependent manner. *Nature Nanotechnology*, 7(6), 383–388.

Cheng, G., Tse, J., Jain, R. K., and Munn, L. L. 2009. Micro-environmental mechanical stress controls tumor spheroid size and morphology by suppressing proliferation and inducing apoptosis in cancer cells. *PLoS One*, 4(2), e4632.

Clevers, H. 2011. The cancer stem cell: Premises, promises and challenges. *Nature Medicine*, 17(3), 313–319.

Cross, S. E., Jin, Y., Rao, J., and Gimzewski, J. K. 2007. Nanomechanical analysis of cells from cancer patients. *Nature Nanotechnology*, 2(12), 780–783.

Deisboeck, T. S., Wang, Z. H., Macklin, P., and Cristini, V. 2011. Multiscale cancer modeling. *Annual Review of Biomedical Engineering*, 13(1), 127–155.

Demou, Z. N. 2010. Gene expression profiles in 3D tumor analogs indicate compressive strain differentially enhances metastatic potential. *Annals of Biomedical Engineering*, 38(11), 3509–3520.

Dong, C., Skalak, R., and Sung, K. L. 1991. Cytoplasmic rheology of passive neutrophils. *Biorheology*, 28, 557–567.

Drasdo, D., and Hoehme, S. 2012. Modeling the impact of granular embedding media, and pulling versus pushing cells on growing cell clones. *New Journal of Physics*, 14(5), 055025.

Drew, D. A., and Segel, L. A. 1971. Averaged equations for two-phase flows. *Studies in Applied Mathematics*, 50(3), 205–231.

Drury, J. L., and Dembo, M. 2001. Aspiration of human neutrophils: Effects of shear thinning and cortical dissipation. *Biophysics Journal*, 81, 3166–3177.

Emens, L. A., Ascierto, P. A., Darcy, P. K. et al. 2017. Cancer immunotherapy: Opportunities and challenges in the rapidly evolving clinical landscape. *European Journal of Cancer*, 81, 116–129.

Fowler, A. C. 1997. *Mathematical Models in the Applied Sciences*. Cambridge, UK: Cambridge University Press.

Frieboes, H. B., Chaplain, M. A. J., Thompson, A. M. et al. 2011. Physical oncology: A bench-to-bedside quantitative and predictive approach. *Cancer Research*, 71(2), 298–302.

Friedl, P., and Gilmour, D. 2009. Collective cell migration in morphogenesis, regeneration and cancer. *Nature Reviews Molecular Cell Biology*, 10(7), 445–457.

Friedl, P., and Wolf, K. 2003. Tumour-cell invasion and migration: Diversity and escape mechanisms. *Nature Review Cancer*, 3(5), 362–374.

Fukumura, D., Duda, D. G., Munn, L. L., and Jain, R. K. 2010. Tumor microvasculature and microenvironment: Novel insights through intravital imaging in pre-clinical models. *Microcirculation*, 17(3), 206–225. doi:10.1111/j.1549-8719.2010.00029.x.

Fuller, R. B. 1961. Tensegrity. *Portfolio Art News Annuals*, (4), 112–127.

Fuller, R. B. 1964. Tensile-integrity structures. United States patent 3063521 A (1962). Patent Link: https://www.google.ch/patents/US3063521.

Galle, J., Aust, G., Schaller, G., Beyer, T., and Drasdo, D. 2006. Individual cell-based models of the spatial-temporal organization of multicellular systems–Achievements and limitations. *Cytometry A*, 69(7), 704–710.

Gang, Z., Qi, Q., Jing, C., and Wang, C. 2009. Measuring microenvironment mechanical stress of rat liver during diethylnitrosamine induced hepatocarcinogenesis by atomic force microscope. *Microscopy Research and Technique*, 72(9), 672–678.

Ginestier, C., Hur, M. H., Charafe-Jauffret, E. et al. 2007. ALDH1 is a marker of normal and malignant human mammary stem cells and a predictor of poor clinical outcome. *Cell Stem Cell*, 1(5), 555–567.

Goel, S., Duda, D. G., Xu, L. et al. 2011. Normalization of the vasculature for treatment of cancer and other diseases. *Physiol Reviews*, 91(3), 1071.

Greenspan, H. P. 1976. On the growth and stability of cell cultures and solid tumors. *Journal of Theoretical Biology*, 56(1), 229–242.

Grzanka, A., Grzanka, D., and Orlikowska, M. 2003. Cytoskeletal reorganization during process of apoptosis induced by cytostatic drugs in K-562 and HL-60 leukemia cell lines. *Cell Stem Cell*, 1(5), 555–567.

Hard, S., Arend, R., Dai, Q. et al. 2017. Implementation and utilization of the molecular tumor board to guide precision medicine. *Oncotarget*, 8(34), 57845. doi:10.18632/oncotarget.18471.

Harrell, M. I., Iritani, B. M., and Ruddell, A. 2007. Tumor-induced sentinel lymph node lymphangiogenesis and increased lymph flow precede melanoma metastasis. *The American Journal of Pathology*, 170, 774–786.

Helmlinger, G., Netti, P. A., Lichtenbeld, H. C., Melder, R. J., and Jain, R. K. 1997. Solid stress inhibits the growth of multicellular tumor spheroids. *Nature Biotechnology*, 15(8), 778–783.

Hessler, J. A., Budor, A., Putchakayala, K. et al. 2005. Atomic force microscopy study of early morphological changes during apoptosis. *Langmuir*, 21(20), 9280–9286. doi:10.1021/la051837g.

Holzapfel, G. A. 2000. Biomechanics of soft tissue. Graz University of Technology, AustriaBiomech Preprint Series., Paper No. 7.

Hou, H. W., Li, Q. S., Lee, G. Y. H., Kumar, A. P., Ong, C. N., and Lim, C. T. 2009. Deformability study of breast cancer cells using microfluidics. *Biomedical Microdevices*, 11(3), 557–564. doi:10.1007/s10544-008-9262-8.

Hoyt, K., Castaneda, B., Zhang, M. et al. 2008. Tissue elasticity properties as biomarkers for prostate cancer. *Cancer Biomarkers*, 4, 213–225.

Iglesias–Garcia, J., Larino–Noia, J., Abdulkader, I., Forteza, J., and Dominguez–Munoz, J. E. 2010. Quantitative endoscopic ultrasound elastography: An accurate method for the differentiation of solid pancreatic masses. *Gastroenterology*, 139(4), 1172–1180.

Ingber, D. E. 1994. Cellular tensegrity and mechanochemical transduction. In V. C. Mow, R. Tran-Son-Tay, F. Guilak and R. M. Hochmuth (Eds.), *Cell Mechanics and Cellular Engineering* (pp. 329–342). New York: Springer.

Jain, R. K. 2013. Normalizing tumor microenvironment to treat cancer: Bench to bedside to biomarkers. *Journal of Clinical Oncology*, 31(17), 2205–2218.

Jain, R. K., Martin, J. D., and Stylianopoulos, T. 2014. The role of mechanical forces in tumor growth and therapy. *Annual Review of Biomedical Engineering*, 16(1), 321–346.

Jain, R. K. 2005. Normalization of tumor vasculature: An emerging concept in antiangiogenic therapy. *Science*, 307(5706), 58.

Kaffas, E. I., Bekah, D., Rui, M., Kumaradas, J. C., and Kolios, M. C. 2013. Investigating longitudinal changes in the mechanical properties of MCF-7 cells exposed to paclitaxol using particle tracking microrheology. *Physics in Medicine & Biology*, 58(4), 923–936.

Kam, Y., Rejniak, K. A., and Anderson, A. R. A. 2012. Cellular modeling of cancer invasion: Integration of in silico and in vitro approaches. *Journal of Cellular Physiology*, 227, 431–438.

Kao, F., Pan, Y., Hsu, R., and Chen, H. 2012. Efficacy verification and microscopic observations of an anticancer peptide, CB1a, on single lung cancer cell. *Biochimica Et Biophysica Acta*, 1818(12), 2927–2935.

Katira, P., Bonnecaze, R. T., and Zaman, M. T. 2013. Modeling the mechanics of cancer: Effect of changes in cellular and extra-cellular mechanical properties. *Frontiers in Oncology*, 3, 145.

Katira, P., Zaman, M. H., and Bonnecaze, R. T. 2012. How changes in cell mechanical properties induce cancerous behavior. *Physical Review Letters*, 108(2), 028103.

Klein, E. A., Yin, L., Kothapalli, D., Castagnino, P. et al. 2009. Cell-cycle control by physiological matrix elasticity and in vivo tissue stiffening. *Current Biology*, 19(18), 1511–1518.

Kocal, G. C., Guven, S., Foygel, K. et al. 2016. Dynamic microenvironment induces phenotypic plasticity of esophageal cancer cells under flow. *Scientific Reports*, 6, 38221.

Kunz-Schughart, L. A., Doetsch, J., Mueller-Klieser, W., and Grebe, K. 2000. Proliferative activity and tumorigenic conversion: Impact on cellular metabolism in 3-D culture. *American Journal of Physiology. Cell Physiology*, 278(4), C765–C780.

Lam, W. A., Rosenbluth, M. J., and Fletcher, D. A. 2007. Chemotherapy exposure increases leukemia cell stiffness. *Blood*, 109(8), 3505–3508.

Lee, G., and Lim, C. T. 2007. Biomechanics approaches to studying human diseases. *Trends in Biotechnology*, 25(3), 111–118.

Liao, D., Sevcencu, C., Yoshida, K., and Gregersen, H. 2006. Viscoelastic properties of isolated rat colon smooth muscle cells. *Cell Biology International*, 30(10), 854–858.

Li, J., Dao, M., Lim, C. T., and Suresh, S. 2005. Spectrin-level modeling of the cytoskeleton and optical tweezers stretching of the erythrocyte. *Biophysical Journal*, 88(5), 3707–3719.

Mani, S. A., Guo, W., Liao, M., Eaton, E. N. et al. 2008. The epithelial-mesenchymal transition generates cells with properties of stem cells. *Cell*, 133(4), 704–715. doi:10.1016/j.cell.2008.03.027.

Marusyk, A., Tabassum, D. P., Altrock, P. M., Almendro, V., Michor, F., and Polyak, K. 2014. Non-cell-autonomous driving of tumour growth supports sub-clonal heterogeneity. *Nature*, 514(7520), 54–58.

McElwain, D. L. S., and Morris, L. E. 1978. Apoptosis as a volume loss mechanism in mathematical models of solid tumor growth. *Mathematical Biosciences*, 39(1–2), 147–157.

Mierke, C. T., Rösel, D., Fabry, B., and Brábek, J. 2008. Contractile forces in tumor cell migration. *European Journal of Cell Biology*, 87, 669–676.

Mills, J. P., Qie, L., Dao, M., Lim, C. T., and Suresh, S. 2004. Nonlinear elastic and viscoelastic deformation of the red blood cell induced by optical tweezers. *Mechanics & Chemistry of Biosystems*, 1, 169–180.

Mitchell, M. J., and King, M. R. 2013. Fluid shear stress sensitizes cancer cells to receptor-mediated apoptosis via trimeric death receptors. *New Journal of Physics*, 15, 015008.

Mysielka, M. E., and Djamgoz, M. B. A. 2004. Cellular mechanism of direct electric filed effects: Galvanotaxis and metastatic disease. *Journal of Cell Science*, 117, 1631–1639.

Nahon, P., Kettaneh, A., Lemoine, M. et al. 2009. Liver stiffness measurement in patients with cirrhosis and hepatocellular carcinoma: A case-control study. *European Journal of Gastroenterology and Hepatology*, 21(2), 214–219.

Ng, C. P., Hinz, B., and Swartz, M. A. 2005. Interstitial fluid flow induces myofibroblast differentiation and collagen alignment in vitro. *Journal of Cell Science*, 118(20), 4731.

Padera, T. P., Stoll, B. R., Tooredman, J. B., Capen, D., Tomaso, E., and Jain, R. K. 2004. Pathology: Cancer cells compress intratumour vessels. *Nature*, 427(6976), 695–695.

Papi, M., Sylla, L., Parasassi, T., Brunelli, R. et al. 2009. Evidence of elastic to plastic transition in the zona pellucida of oocytes using atomic force spectroscopy. *Applied Physics Letters*, 94(15), 153902.

Pedersen, J. A., Boschetti, F., and Swartz, M. A. 2007. Effects of extracellular fiber architecture on cell membrane shear stress in a 3D fibrous matrix. *Journal of Biomechanics*, 40(7), 1484–1492.

Phipps, C., and Kohandel, M. 2011. Mathematical model of the effect of interstitial fluid pressure on angiogenic behavior in solid tumors. *Computational and Mathematical Methods in Medicine*, Volume 2011(Article ID 843765), 9 pages.

Prasetyanti, P., and Medema, J. P. 2017. Intra-tumor heterogeneity from a cancer stem cell perspective. *Molecular Cancer*, 16(41), 1–9.

Radmacher, M., Fritz, M., Kacher, C. M., Cleveland, J. P., and Hansma, P. K. 1996. Measuring the viscoelastic properties of human platelets with the atomic force microscope. *Biophysical Journal*, 70(1), 556–567.

Randel L. Swanson, II, DO, "Biotensengrity: A unifying theory of biological architecture with applications to osteopathic practice, education and research-A review and analysis," The J of the American Osteopathic Association, Jan 2013, Vol 113, No 1, P 34-52. Link: http://studylib.net/doc/8915958/biotensegrity--a-unifying-theory-of-biological-architecture.

Rejniak, K. A., and McCawley, L. J. 2010. Current trends in mathematical modeling of tumor–microenvironment interactions: A survey of tools and applications. *Experimental Biology and Medicine*, 235(4), 411–423.

Rodriguez-torres, M., and Allan, A. L. 2016. Aldehyde dehydrogenase as a marker and functional mediator of metastasis in solid tumors. *Clinical and Experimental Metastasis*, 33(1), 97–113.

Roose, T., Netti, P. A., Munn, L. L., Boucher, Y., and Jain, R. K. 2003. Solid stress generated by spheroid growth estimated using a linear poroelasticity model. *Microvascular Research*, 66(3), 204–212.

Rosenbluth, M. J., Lam, W. A., and Fletcher, D. A. 2006. Force microscopy of nonadherent cells: A comparison of leukemia cell deformability. *Biophysical Journal*, 90(8), 2994–3003.

Ruddell, A., Harrell, M., Minoshima, S. et al. 2008. Dynamic contrast-enhanced magnetic resonance imaging of tumor-induced lymph flow. *Neoplasia*, 10(7), 706–IN4.

Sabass, B., Gardel, M. L., Waterman, C. M., and Schwarz, U. S. 2008. High resolution traction force microscopy based on experimental and computational advances. *Biophysical Journal*, 94(1), 207–220.

Sakuma, S., Kuroda, K., Tsai, C. D., Fukui, W., Arai, F., and Kaneko, M. 2014. Red blood cell fatigue evaluation based on the close-encountering point between extensibility and recoverability. *Lab on a Chip*, 14(6), 1135–1141.

Sanga, S., Frieboes, H. B., Zheng, X., Gatenby, R., Bearer, E. L., and Cristini, V. 2007. Predictive oncology: A review of multidisciplinary, multiscale in silico modeling linking phenotype, morphology and growth. *Neuroimage*, 37, S120–S134.

Sarntinoranont, M., Rooney, F., and Ferrari, M. 2003. Interstitial stress and fluid pressure within a growing tumor. *Annals of Biomedical Engineering*, 31(3), 327–335.

Schmid-Schonbein, G. W., Kosawada, T., Skalak, R., and Chien, S. 1995. Membrane model of endothelial cells and leukocytes. A proposal for the origin of a cortical stress. *Journal of Biomechanical Engineering*, 117(2), 171–178.

Shao, Z. M., Nguyen, M., and Barsky, S. H. 2000. Human breast carcinoma desmoplasia is PDGF initiated. *Oncogene*, 19(38), 4337–4345.

Sharma, S., Santiskulvong, C., Bentolila, L. A., Rao, J., Dorigo, O., and Gimzewski, J. K. 2012. Correlative nanomechanical profiling with super-resolution F-actin imaging reveals novel insights into mechanisms of cisplatin resistance in ovarian cancer cells. *Nanomedicine: Nanotechnology, Biology and Medicine*, 8(5), 757–766.

Shieh, A. C. 2011. Biomechanical forces shape the tumor microenvironment. *Annals of Biomedical Engineering*, 39(5), 1379–1389.

Sottoriva, A., Verhoeff, J. J. C., Borovski, T. et al. 2010. Cancer stem cell tumor model reveals invasive morphology and increased phenotypical heterogeneity. *Cancer Research*, 70(1), 46.

Stolarska, M. A., Kim, Y., and Othmer, H. G. 2009. Multi-scale models of cell and tissue dynamics. *Philosophical Transactions of the Royal Society A: Mathematical, Physical and Engineering Sciences*, 367, 3525–3553.

Stylianopoulos, T., Martin, J. D., Chauhan, V. P. et al. 2012. Causes, consequences, and remedies for growth-induced solid stress in murine and human tumors. *Proceedings of the National Academy of Sciences of the United States of America*, 109(38), 15101.

Sun, Y., Wan, K. T., Roberts, K. P., Bischof, J. C., and Nelson, B. J. 2003. Mechanical property characterization of mouse zona pellucida. *IEEE Transaction on Nanobioscience*, 2(4), 279–286.

Suresh, S. 2007. Biomechanics and biophysics of cancer cells. *Acta Biomaterialia*, 3(4), 413–438.

Sutherland, R. M. 1988. Cell and environment interactions in tumor microregions: The multicell spheroid model. *Science*, 240(4849), 177–184.

Swanson, R. L. 2013. Biotensengrity: A unifying theory of biological architecture with applications to osteopathic practice, education and research-A review and analysis. *The Journal of the American Osteopathic Association*, 113(1), 34–52.

Swartz, M. A., and Fleury, M. E. 2007. Interstitial flow and its effects in soft tissues. *Annual Review of Biomedical Engineering*, 9, 229–256.

Swartz, M. A., and Lund, A. W. 2012. Lymphatic and interstitial flow in the tumour microenvironment: Linking mechanobiology with immunity. *Nature Reviews Cancer*, 12(3), 210–219.

Tadeo, I., Bernegal, A. P., Escudero, L. M., Álvaro, T., and Noguera, R. 2014. Biotensegrity of the extracellular matrix: Physiology, dynamic mechanical balance, and implications in oncology and mechanotherapy. *Frontiers in Oncology*, 4(39), 1–10.

Thoumine, O., and Ott, A. 1997. Time scale dependent viscoelastic and contractile regimes in fibroblasts probed by microplate manipulation. *Journal of Cell Science*, 110, 2109–2116.

Thoumine, O., Ott, A., Cardoso, O., and Meister, J. 1999. Microplates: A new tool for manipulation and mechanical perturbation of individual cells. *Journal of Biochemical and Biophysical Methods*, 39(1), 47–62.

Tsai, C. D., Sakuma, S., Arai, F., and Kaneko, M. 2014. A new dimensionless index for evaluating cell stiffness-based deformability in microchannel. *IEEE Transactions on Biomedical Engineering*, 61(4), 1187–1195.

Tse, J. M., Cheng, G., Tyrrell, J. A. et al. 2012. Mechanical compression drives cancer cells toward invasive phenotype. *Proceedings of the National Academy of Sciences of the United States of America*, 109(3), 911.

Tseng, Y., Lee, J. S., Kole, T. P., Jiang, I., and Wirtz, D. 2004. Micro-organization and visco-elasticity of the interphase nucleus revealed by particle nanotracking. *Journal of Cell Science*, 117(10), 2159–2167.

Van Marck, V., Stove, C., Van, D. B. et al. 2005. P-cadherin promotes cell-cell adhesion and counteracts invasion in human melanoma. *Cancer Research*, 65(19), 8774.

Voutouri, C., Mpekris, F., Papageorgis, P., Odysseos, A. D., and Stylianopoulos, T. 2014. Role of constitutive behavior and tumor-host mechanical interactions in the state of stress and growth of solid tumors. *PLoS One*, 9(8), e104717.

Waclaw, B., Bozic, I., Pittman, M. E., Hruban, R. H., Vogelstein, B., and Nowak, M. A. 2015. A spatial model predicts that dispersal and cell turnover limit intratumour heterogeneity. *Nature*, 525(7568), 261–264.

Wang, N., Naruse, K., Stamenovic, D., Fredberg, J. J. et al. 2001. Mechanical behavior in living cells consistent with the tensegrity model. *Proceedings of the National Academy of Sciences of the United States of America*, 98(14), 7765–7770.

Ward, J. P., and King, J. R. 1997. Mathematical modelling of avascular-tumour growth. *IMA Journal of the Mathematical Applied Medicine and Biology*, 14(1), 39–69.

Ward, J. P., and King, J. R. 1999. Mathematical modelling of avascular-tumour growth II: Modeling growth saturation. *IMA Journal of the Mathematical Applied Medicine and Biology*, 16(2), 171–211.

Weiss, L. 1992. Comments on hematogenous metastatic patterns in humans as revealed by autopsy. *Clinical and Experimental Metastasis*, 10(3), 191–199. doi:10.1007/BF00132751.

Wirtz, D., Konstantopoulos, K., and Searson, P. C. 2011. The physics of cancer: The role of physical interactions and mechanical forces in metastasis. *Nature Reviews Cancer*, 11(7), 512–522.

Xu, W., Mezencev, R., Kim, B., Wang, L., McDonald, J., and Sulchek, T. 2012. Cell stiffness is a biomarker of the metastatic potential of ovarian cancer cells. *PLoS One*, 7(10), 46609.

Zaman, M. H., Trapani, L. M., Sieminski, A. L., MacKellar, D. et al. 2006. Migration of tumor cells in 3D matrices is governed by matrix stiffness along with cell-matrix adhesion and proteolysis. *Proceedings of the National Academy of Sciences of the United States of America*, 103(29), 10889.

Zhao, G., Cui, J., Qin, Q. et al. 2010. Mechanical stiffness of liver tissues in relation to integrin β1 expression may influence the development of hepatic cirrhosis and hepatocellular carcinoma. *Journal of Surgical Oncology*, 102(5), 482–489.

5

Biomechanical Modeling Applications in Image-Guided Radiotherapy

Michael Velec and
Kristy K. Brock

5.1 Introduction

As the demands for improved accuracy in radiotherapy increase, so does the need for better accounting of the dynamic nature of patients. Radiotherapy is a primary cancer treatment modality that benefits from integrating information acquired with multi-modality and serial images at all stages of the treatment process. This wealth of imaging information creates a demand for deformable image registration (DIR) to model anatomical changes caused by physiological processes (e.g., breathing motion, rectal and bladder filling, and peristalsis), changes in patient pose (e.g., patient positioning at each treatment and between imaging sessions), and the responses of tissues to radiation. Unlike many other algorithms that are available, biomechanical-based DIR algorithms model the physical properties of tissues that govern many of these sources of motion.

Radiotherapy plans are often designed on a single-static computed tomography (CT) image of the patient acquired before treatment. To account for geometric uncertainties in the treatment processes, radiation dose is planned to tumor volumes plus an additional safety margin at a cost of increased normal tissue irradiation. During treatment delivered over multiple fractions, spanning days-to-weeks, daily image guidance is employed to aid in targeting tumors and improve confidence that the planned dose is accurately delivered to the patients. Motion and deformation during delivery, however, can cause deviations in the intended doses exceeding 5% in magnitude. Reducing the uncertainty in these processes through the application of DIR can potentially improve the therapeutic ratio. This chapter initially focuses on briefly describing biomechanical-based DIR techniques developed in several anatomic sites. The second half illustrates how the application of biomechanical DIR can reduce geometric and dosimetric uncertainties in radiotherapy, improve the design of treatment, and augment our understanding of response in several clinical scenarios.

5.2 Biomechanical Deformable Models

Biomechanical modeling has been investigated for medical applications ranging from surgery to radiology. This section highlights biomechanical DIR techniques reported in the literature specifically intended for applications in radiotherapy. Because radiotherapy requires motion models to be as realistic and accurate as possible, there is motivation to develop DIR algorithms that are physiologically plausible. This is the foundation for the development of biomechanical models that incorporate the underlying principles governing soft tissue deformations. Although other approaches, such as intensity-based DIR algorithms, for example, may also be evaluated by their physiological plausibility that they are not physically motivated models.

DIR development and validation is an area of active research; therefore, it is important to appreciate that the accuracy can vary with implementations, and the accuracy requirements vary by application. Validation of DIR remains a challenging task as the ground truth may not be accurately known. Methods can include performing registration on physical or mathematical phantoms with known transformations. In patient data, anatomic regions of interest and landmark points defined on the primary image can be propagated using the deformation vector field (DVF) and compared to their actual positions defined in the secondary images. A common accuracy metric is the Dice similarity coefficient (DSC), which quantifies the degree of overlap between contours deformed from the primary image (A′) to the actual contour in the secondary image (B) using the equation $DSC = 2(A' \cap B)/(A' + B)$. A DSC value of 0 indicates that there is no overlap of the contours, whereas a value of 1 indicates perfect agreement. DSC must be interpreted with consideration of the baseline uncertainty in clinicians generating the contours, whereas small or irregularly shaped contours often have inherently lower DSC than larger volumes. Contours can alternatively be compared after conversion to a series of points and calculating the distances between points in the images. Common metrics reported include the mean distance-to-agreement of the vector displacements or the maximum distance-to-agreement (sometimes referred to as the Hausdorff distance). The target registration error (TRE) is also often used to quantify the accuracy at point landmarks that are not used to drive the deformation. The TRE is the average 3D vector magnitude between the landmark points A′ and B, or the distances between these can be reported as per the three anatomical planes. A value of 0 mm for these distance-based measures would represent perfect registration at the contours or landmarks of interest.

5.2.1 Implementation Examples in the Literature

5.2.1.1 Finite Element Method-Based Methods

As part of the typical radiotherapy process contours of tumors and normal tissues are delineated on the primary image dataset, which is often the planning CT dataset. These segmentations can be used to construct single or multiorgan 3D mesh models of the patient forming the basis for biomechanical models using the finite element method (FEM). Material properties are assigned to different tissues and boundary conditions that are typically established at tissue surfaces defined on primary and secondary images. By the use of biomechanical FEM model based on the primary image and the boundary conditions, a system of constitutive equations is generated and solved using finite element analysis (FEA). The displacements for the nodes comprising all the tissues in the biomechanical model are thus solved resulting in a displacement map, or DVF, which can then be used to perform DIR. As manual tissue delineation is often only possible when there is visible intensity gradient in the images, deriving boundary conditions from delineated surfaces avoids the difficulty in defining correspondences in regions away from tissue interfaces (e.g., within homogenous tissues).

Using this approach, researchers at William Beaumont Hospital, Royal Oak, Michigan, first proposed the use of FEM to accumulate radiotherapy doses calculated on serial CT images acquired during the treatment course to account for day-to-day organ motion (Yan et al. 1999). A hollow hexahedral element mesh model of the rectal wall, an organ at risk, was created from the planning CT contours of a prostate radiotherapy patient. As patient-specific values are not known, linear elastic material property values for Poisson's ratio (ν) and Young's Modulus (E) were applied from the literature. Boundary conditions were defined simply by manually identifying point-to-point correspondences of the outer and inner rectum surfaces between the planning CT and repeat CT scans. Manually selecting limited landmarks on organ surfaces can, however, be prone to significant uncertainties. This process was subsequently refined using an optimization scheme based on the principle of energy minimization to establish automated boundary conditions over the entire organ surface and reduce these uncertainties within 1 mm (Liang and Yan 2003). This technique

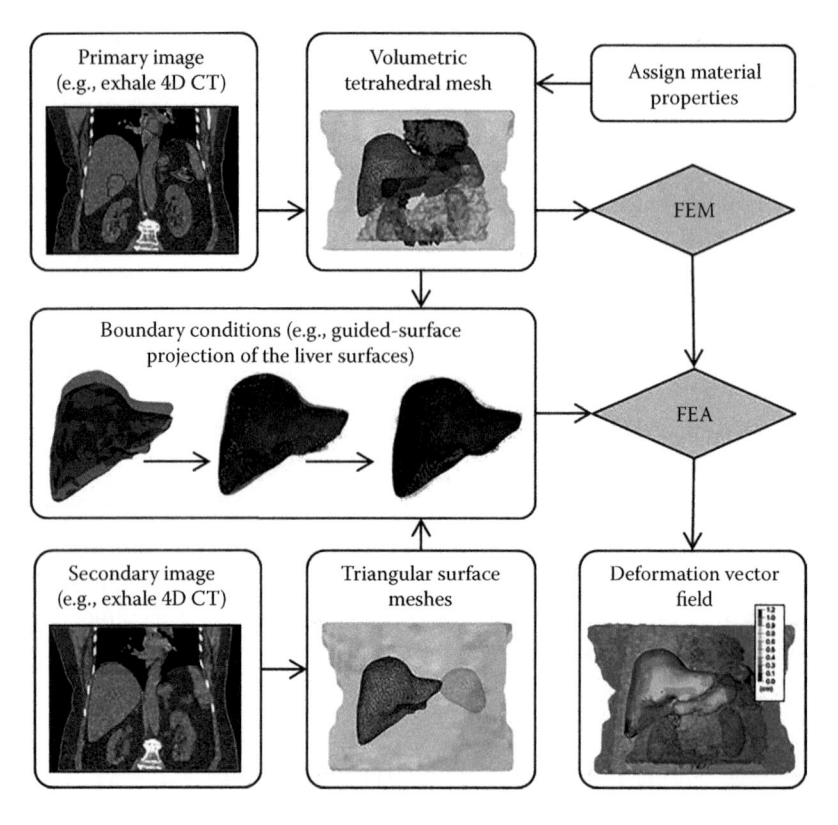

FIGURE 5.1   Schema of multiorgan, biomechanical model-based deformable image registration (Morfeus).

was expanded to include modeling of the bladder and prostate gland with each tissue solved independently (Wu et al. 2006).

A multiorgan biomechanical model-based DIR algorithm, Morfeus, was developed at the Princess Margaret Cancer Centre, Toronto, Ontario, Canada (Brock et al. 2005b). The framework is summarized in Figure 5.1. The tetrahedral models are typically configured with tied constraints at tissue interfaces (e.g., tumor and the tumor-bearing organ, or organ and the external body), except for the lungs and chest wall interface where a frictionless surface-based contact model is applied to permit sliding (Figure 5.2) (Al-Mayah et al. 2009b). Linear elastic material properties are adapted from the literature and optimized for registration accuracy (Brock et al. 2005b). Boundary conditions are determined for a subset of visible organs that are contoured on both image datasets. These organ surfaces undergo explicit surface deformation using an automated guided-surface projection technique without the requirement to manually define node-to-node correspondences between surfaces. This DIR algorithm has been evaluated in the head and neck, thorax, abdomen, and pelvis, with TRE reported from 2–4 mm, which is similar to the average voxel resolutions (Al-Mayah et al. 2009a;

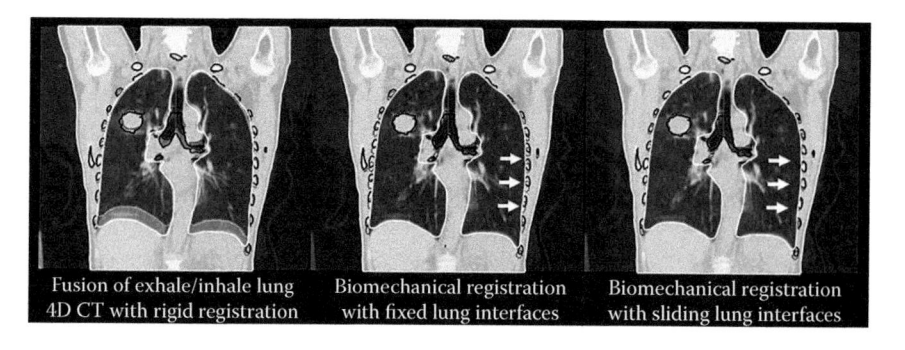

| Fusion of exhale/inhale lung 4D CT with rigid registration | Biomechanical registration with fixed lung interfaces | Biomechanical registration with sliding lung interfaces |

FIGURE 5.2   Lung 4D CT example of biomechanical registration implemented in a commercial treatment-planning system. Incorporating sliding interfaces on the lung meshes (white contours) improves alignment of adjacent structures (black contours), in particular the ribs (arrows).

Al-Mayah et al. 2010; Al-Mayah et al. 2011; Brock et al. 2005b; Brock et al. 2008b; Hensel et al. 2007; Voroney et al. 2006).

The diverse techniques used to establish boundary conditions are among the most significant variations between FEM implementations. Crouch et al. (2007) reported on a 3D technique for the prostate where boundary conditions are created by adapting a medial shape model to binary mask images of the prostate using bicubic Bézier splines. In this implementation, a hexahedral prostate mesh is created from the shape model and solved with FEA. The prostate TRE calculated from implanted brachytherapy seeds was 5.2 mm compared to a TRE of 10.5 mm with rigid registration and an image slice thickness of 5 mm. There has been particular interest in using FEM models to perform DIR of prostate images acquired with and without endorectal coils, which induces substantial deformation of the gland. Alterovitz et al. (2006) reported on the use of FEM to perform 2D registration on MR images of the prostate. The model consists of 2D triangular elements and treats the values for Young's modulus and the external forces as uncertain parameters, which are found using an optimization solution. The DIR accuracy for 10 cases resulted in an average prostate DSC of approximately 98%. The use of 2D DIR is potentially an efficient method for clinical implementation provided the most relevant 2D slices requiring registration can be identified. Bharatha et al. (2001) developed a linear elastic full 3D biomechanical prostate model where heterogeneous material properties are assigned to the different intraprostatic zones. The boundary conditions are automatically produced using an active-surface algorithm, which iteratively matches the prostate segmentations. In more than 10 prostate patients, the DSC of the intraprostatic regions significantly improved from 0.59–0.81 with rigid registration to 0.76–0.94 with DIR.

The concept of inflating organs to drive the deformation by directly simulating the mechanical forces acting on the organs has also been explored. Boubaker et al. (2009) used a hyperelastic FEM model of the male pelvis based on CT imaging. Frictionless contact surfaces were applied to permit sliding between the prostate, bladder, rectum, and pelvic bones. Internal pressures of the bladder and rectum, hollow structures comprising only of triangular surface meshes, are then uniformly and incrementally applied to the internal surfaces to serve as boundary conditions for the multiorgan FEM model.

The optimal solution is reached when the prostate is deformed into the correct secondary position, resulting in an 8% relative error in the prostate. Similar approaches have demonstrated that the bladder can be accurately deformed with DSC values exceeding 0.87 (Chai et al. 2012). Werner et al. modeled sliding of the lungs by progressively inflating the lung volumes from the exhale geometry on 4D CT (Werner et al. 2009). A uniform negative pressure is initially applied to the lungs. Intraplural pressure is then gradually increased to force the lungs to expand while sliding against the chest wall using frictionless contact surfaces, except for the fixed regions areas where bronchi enter the lungs. The expansion is limited by the lung's fixed geometry on the secondary inhale 4D CT. Compared to rigid registration with a TRE of 6.6 ± 5.2 mm on 12 patients, the TRE using the biomechanical model was 3.3 ± 2.1 mm.

5.2.1.2 Finite Element Method Model Parameter Sensitivities

The desire to account for organ motion using physically based models for DIR is challenging in practice as the actual *in vivo* material properties of tissues are in most scenarios not actually known. Therefore, the applied material properties (i.e., v, E) are typically optimized either on a test cohort of data or included in the optimization scheme on a patient-specific basis. The degree of sensitivity in material properties on DIR accuracy varies among implementations. Chi et al. (2006) varied anisotropic properties for a linear elastic biomechanical FEM model of the pelvis consisting of a solid prostate gland, and hollow rectum and bladder organs. A 30% material uncertainty resulted in displacement differences up to 1.3 mm for the hollow organs and 4.5 mm for the prostate; however, most differences were submillimeter. Tehrani compared FEM sensitivity of material properties in the lung for three elasticity models including an isotropic linear elastic model, and neo–Hookean or Mooney–Rivlin hyper-elastic models (Tehrani et al. 2015). Across the eight patients evaluated, the optimized material properties were both patient- and model-specific. Overall the mean 3D error in the predicted tumor position was 4.0 mm using the linear elastic, whereas both hyperelastic models had mean errors of 2.3 mm. Li et al. (2013) compared thoracic FEM models with uniform or heterogeneous elasticity values applied across the lung parenchyma. An intensity based block-matching technique was used to derive the boundary conditions at the lung surfaces between exhale and inhale 4D CT. Element-specific Young's modulus was estimated by solving an optimization problem using a quasi-Newton approach to minimize the differences between FEM results and those predicted by the intensity based step. Therefore, the distribution of E varies for each patient, and spatially including the tumor area without an explicit delineation. Compared to the uniform model, the heterogeneous model reduces the TRE from 3.1 ± 2.4 mm to 1.3 ± 1.0 mm. In the above-mentioned examples, plausible material properties were applied resulting in DIR accuracy on par with the image voxel resolution without the requirement for quantifying *in vivo* parameters.

5.2.1.3 Hybrid and Alternative Approaches

To combine the potential advantages of biomechanical and intensity-based algorithms, sequentially applied DIRs have been proposed. Samavati investigated this approach in lung 4D CT by applying a FEM-based DIR with contact surfaces to account for the sliding motion followed by a B-spline algorithm to refine the registration through maximizing the sum of absolute image-intensity differences (Samavati et al. 2015). The TRE over the 31 patients studied improved from 3.1 ± 1.9 mm using the biomechanical technique alone to 1.5 ± 1.4 using the hybrid method. Improving the accuracy of intensity-based algorithms in low-contrast regions has also been performed by applying a subsequent biomechanical DIR for prostate and lung imaging (Zhong et al. 2012; Zhong et al. 2015). In another implementation for lung 4D CT, images were automatically segmented into binary masks based on high-contrast features and registered using a Demons intensity-based DIR algorithm. A uniform tetrahedral FEM model of the entire image is generated without the effort and uncertainty associated with manual segmentation of individual organs. The model, encompassing high- and low-contrast regions is deformed using the Demons-derived displacements as the boundary conditions. On a mathematical phantom of lung 4D CT, the FEM-based correction technique reduced the maximum error from 12 to 4 mm and average errors from 1.7 to 1.1 mm owing to the improved regularization in low-contrast regions.

In the same manner of using intensity based DIR to establish the boundary conditions, Cazoulat et al. (2016) generated binary images of the lung vessel trees using histogram thresholding followed by a Demons intensity based DIR to determine the vessel centerline displacements. An FEM-based registration is performed using a contact-surface technique at the lung surfaces plus the internal vessel displacements as the boundary conditions. The technique was tested on six patients with exhale-phase 4D CT images acquired for treatment planning and three weeks into a radiotherapy, which exhibited tumor volume reductions of 12%–36%. The TRE was 5.8 ± 2.9 mm with rigid registration, 3.4 ± 2.9 mm using the biomechanical DIR alone, and 1.6 ± 1.3 mm with the additional vessel-based boundary conditions with particular improvement noted in the tumor-bearing lung region. Because the accuracy of the biomechanical model-based registration was affected by large anatomical changes not solely caused by biomechanical processes (i.e., tumor regression), the addition of the intensity based boundary conditions improved the accuracy.

In an example not relying on FEM, Neylon et al. (2015) used a computer-graphics approach to simulate tissue biomechanics in the head and neck. Individual bones are modeled as rigid bodies, whereas soft tissues are modeled as a series of connected mass elements with mass-spring dampers and elastic material properties. The motion is initiated by prescribing rigid motion of each bone that causes soft tissues to undergo elastic deformations due to tensile and shear spring forces and damping forces. To model tumor regression, a change that is not governed strictly by biomechanics, the rest lengths of the connections within the tumor were reduced to initiate forces that reflect volume decreases. For 10 radiotherapy patients, DIR of the planning CT to CT acquired on the final week treatment resulted in an increased image similarity with correlation coefficient increasing on an average from 0.844 ± 0.136 to 0.960 ± 0.022. Li et al. (2016) similarly avoids the FEM requirements of mesh model building and meshing nonoverlapping tissues by using a total Lagrangian explicit dynamics algorithm for whole body image registration. Mechanical properties are assigned to the meshless computational grid using a fuzzy tissue classification of the image intensities and boundary conditions that are applied to individual vertebrae based on their individual rigid registration. Compared to nonlinear FEM-based DIR of whole body images, this approach resulted in nodal displacement differences of less than 1 mm on more than 99% of the nodes.

5.2.2 Accuracy Comparisons to Other Deformable Image Registration Algorithms

The numerous DIR variations reported in the literature have motivated studies to directly compare algorithm results on common datasets. These efforts may aid clinicians in choosing suitable algorithms for specific anatomical sites and applications. In a multiinstitution study investigating DIR accuracy on patient images, results were submitted from academic institutions and medical device manufactures (Brock 2010). More than 20 algorithms produced results for 4D CT images and the only biomechanical DIR included in the study had TRE within 1.2 mm of the best performing algorithm in the lungs and within 0.7 mm of the best performing algorithm in the liver. On liver CT-MR and repeat prostate MR data, the biomechanical DIR had TREs of 3.9 and 2.3 mm, respectively, which was the best of the only three algorithms that provided results for these datasets. The biomechanical DIR was accurate within the image-voxel resolutions and demonstrated its versatility in multimodality registration as the integration of non-CT images into radiotherapy workflows increases.

A deformable 3D dosimeter developed by Juang et al. (2003) was developed to quantitatively assess DIR accuracy over all voxels in the images, in contrast to patient data where it is often only possible to sparsely identify anatomic landmarks for evaluation. The radiochromic material undergoes a linear dose response when irradiated and the delivered dose distribution is measured in 3D using an optical CT platform. The phantom was imaged and irradiated in a deformed state and subsequently imaged in its undeformed state following irradiation. The exact same data was used in two separate studies evaluating DIR. The first applied a commercially available intensity based algorithm DIR using a B-spline deformation and mutual information-similarity metric. CT images were registered with submillimeter accuracy at the dosimeter's surface. However, the predicted irradiated field edges within the centre of the dosimeter had errors of approximately 4 mm on average up to a maximum of 14 mm. Only 60% of the voxels in the DIR-predicted distribution passed acceptance criteria commonly used in the clinical setting ($\gamma_{3\%/3\ mm}$), which is only marginally better than rigid registration with $\gamma_{3\%/3\ mm}$ of 58%. The second study applied a FEM-based biomechanical algorithm that resulted in predicted irradiated field edges with an accuracy of 1 mm on average with maximum errors of 3 mm, results that are comparable to other geometric and dosimetric uncertainties typical for this end-to-end test (Figure 5.3) (Velec et al. 2015). More than 90% of the voxels in the DIR-predicted distribution passed acceptance criteria that are commonly used in the clinical setting ($\gamma_{3\%/3\ mm}$). These results show promise of biomechanical DIR to deform doses in the presence of large deformation magnitudes and in structures with homogenous appearances on CT. The intensity-based DIR, which was otherwise shown to perform well on feature-rich patient imaging in other evaluations, likely struggled with the uniform appearance of dosimeter on CT. More caution may therefore be required when applying intensity-based DIR to images, modalities, or structures for which it has not yet been validated. In comparison, biomechanical-based DIR functions independently of image intensities that save for the contouring of the structural surfaces.

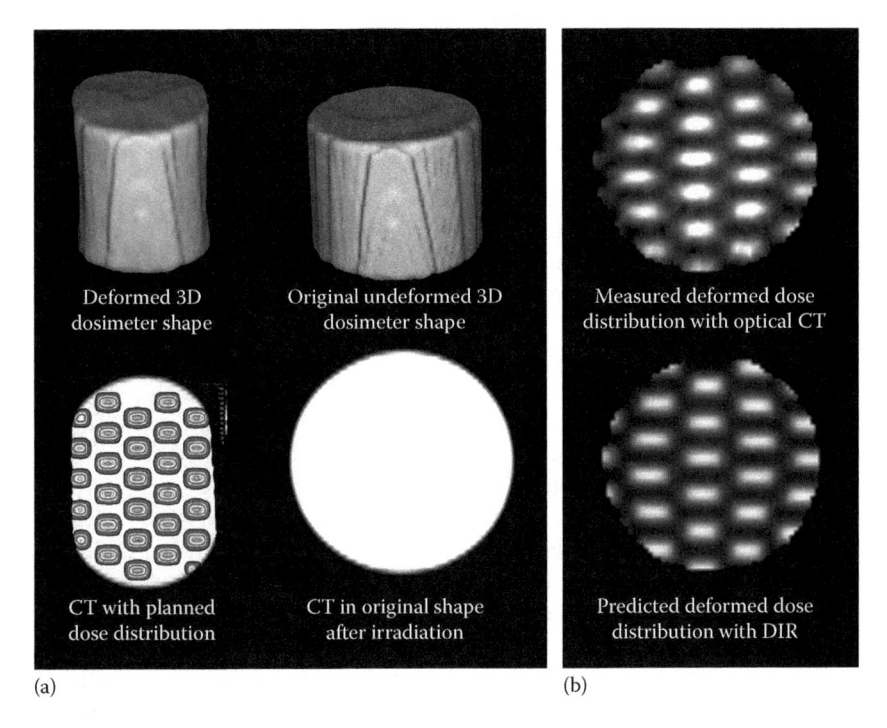

| Deformed 3D dosimeter shape | Original undeformed 3D dosimeter shape | Measured deformed dose distribution with optical CT |
| CT with planned dose distribution | CT in original shape after irradiation | Predicted deformed dose distribution with DIR |

(a)   (b)

FIGURE 5.3   A deformable radiochromic dosimeter before and after a temporary deformation (a). Measured dose from optical CT is compared to the predicted dose reconstructed with biomechanical DIR of the CT images (b). (Adapted from Velec, M. et al., *Pract. Radiat. Oncol.*, 5, e401–e408, 2015. With permission.)

Therefore, its accuracy can be reasonably assumed to be valid for similar geometries and deformations, regardless of the contrast in the image.

5.3 Applications in Radiotherapy

The following sections illustrate the applications of biomechanical model-based DIR toward improving the spatial and dosimetric accuracy of radiotherapy, and aiding the interpretation of patient responses to treatment. The published examples in the literature demonstrate utility in a range of clinical scenarios that span the patient's trajectory from diagnosis, planning, and delivery of radiotherapy, and to posttreatment follow-up. Although the majority of the studies have focused on retrospective application, biomechanical DIR algorithms have been recently implemented in a commercial treatment-planning system, which should facilitate prospective use in patients. Currently, there is no consensus on how to commission DIR algorithms in general and the level of accuracy that is required for each application.

Therefore, further progress toward validation will allow for more complex techniques, such as dose accumulation, to become implemented in clinics.

5.3.1 Quality Assurance of Deformable Image Registration

The difficulty in DIR validation on patient images has led to the proposed application of using FEM-based biomechanical models to evaluate the DVF results of other algorithms. In this manner, the results from FEM provide a mechanically plausible *gold standard*. Zhong et al. (2007) proposed an approach that uses a biomechanical FEM model as the frame of reference in order to assess misregistration causes by DIR inaccuracies, low image quality, and artifacts. Assuming that the internal strain energy of a deforming tissue must be matched by external forces, quantifying unbalanced energies at the voxel level allows spatial assessment of the DVF for violations of continuum mechanics principles. These local inconsistencies correlated well with the local TRE on lung 4D CT images and are implied to affect the displacement errors in the DVF (Li et al. 2013). Thus, the technique can be applied in settings where the ground truth deformation is not known and it may not be possible to directly quantify DVF error. The utility of this approach is that it provides automated, 3D, objective evaluation on patient-specific basis on clinical images without the need to identify anatomical correspondence (e.g., point landmarks), which is labor intensive and not possible on all image modalities. An extension of the technique integrates information of the local planned dose gradients to estimate the errors in doses deformed through the DIR without the explicit knowledge of the ground truth (Zhong et al. 2008).

5.3.2 Treatment Planning

Resolving the spatiotemporal differences caused by organ motion and deformation may enable improved assessment and integration of the tumor information provided by each image or modality. In radiotherapy, a model of the patient is typically represented by a single volumetric image in the treatment-planning system. This is often a CT scan acquired with the patient immobilized in a stable position intended to be reproduced during subsequent treatment delivery. Integrating information from additional images (e.g., contrast-enhanced CT, MRI, PET) is often needed to aid in defining the boundary of the tumor and quantify physiological motion (e.g., respiration) to design patient-specific planning target volume (PTV) margins. Currently, additional image information is often mapped to the planning CT through rigid registration. The limited degrees-of-freedom (DOF) through which rigid registration can map this information, is a source of uncertainty. At the treatment-planning stage, it is particularly important to achieve a high degree of accuracy to prevent the propagation of systematic errors throughout the course of treatment.

5.3.2.1 Target Delineation

Although treatment is typically planned on CT images, additional image datasets from modalities such as MRI are increasingly being relied on to define the tumor due to superior soft tissue contrast and functional information. These MRI scans may be acquired

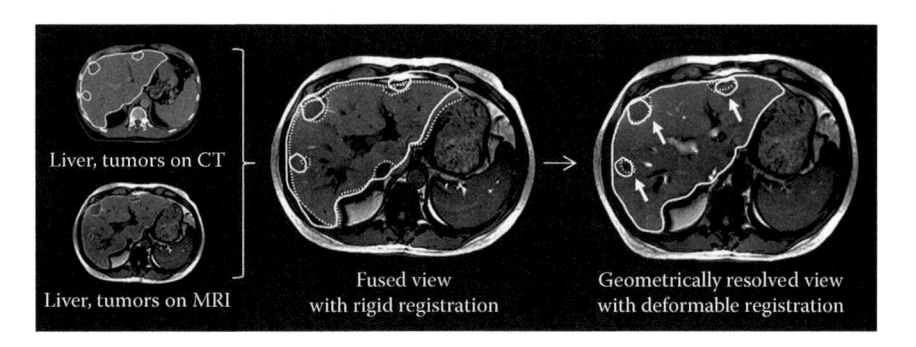

FIGURE 5.4 Biomechanical model-based registration of the liver is used to generate a geometrically resolved view. The spatial integrity of tumors caused by differences in representation on multimodality images (white arrows) is preserved. Solid contours define regions on the CT and dashed contours define regions on MRI.

in the radiotherapy department on the same day as the CT, permitting the patient's position to be reproduced, or they may be acquired beforehand in the diagnostic radiology department. Soft tissue deformations occurring between the planning CT and MRI or other modalities potentially limit the accuracy of integrating all the information into one model of the patient, if rigid registration is used.

By applying the biomechanical DIR, as demonstrated in Figure 5.4, geometric differences caused by organ deformation can be resolved while preserving the spatial integrity of information in the form of image intensity (e.g., variable tumor representation on serial or multimodal imaging) (Brock et al. 2006). Geometrically resolved images can be generated using a ray-driven approach where the intensity values for the deformed image are found from the corresponding voxel on the undeformed image using the DVF (Samavati et al. 2015; Brock et al. 2005a). With biomechanical model-based DIR, the mapping is independent of neighboring voxel intensities in the initial image that can permit sheering (i.e., lung sliding) in the final deformed image.

Using the radiotherapy planning for liver tumors as a clinical example, MRI is often rigidly registered to the planning CT to aid in delineation of the target. Liver tumors visualized on MRI can differ in shape, size, and number of lesions as compared to CT; thus, providing complementary information. These can pose challenges for accurate image registration, as the inherent intensity differences between modalities or use of intravenous contrast. The liver boundary is both salient and similarly represented on CT and MR, and therefore often used as a surrogate for registration. In a study of 26 patients, biomechanical DIR was performed between the liver volumes on the planning CT and MRI, both acquired at exhale breath-hold on the same day (Voroney et al. 2006). The geometrically resolved CT thus accounted for any changes in liver position relative to bony anatomy and deformation allowing for a comparison of tumors delineated on each image. Following DIR, a median (range) of 95% of the tumor surfaces differed by 9 mm (4–52 mm) between the geometrically resolved CT and MR, owing to inherent imaging differences, and tumor representation, between the modalities. Most important, the concordance volume of tumors between images improved from 65% on average using rigid-vertebral body registration to 73% with deformable registration. Eliminating differences in tumor definition due to liver deformation and therefore increasing the geometric accuracy means, in practice, that a smaller volume of healthy liver would require radiation. The technique additionally provides a mechanism to accurately delineate and map tumor defined on MR for patients who have a contraindication to intravenous contrast used for planning CT acquisition (Brock et al. 2006).

In prostate radiotherapy, there is a clinical need to fuse planning CT images of the pelvis to MRI acquired with an inflated endorectal coil (Hensel et al. 2007). The later image has enhanced internal structural and boundary details of the prostate gland needed for radiotherapy planning and can also guide biopsy during diagnostic workup (Menard et al. 2015). A limitation of endorectal coils is the substantial pelvic deformation that it causes at planning. Because the coils are not used during treatment delivery, spanning 30 or more daily fractions, other approaches are required to minimize the systematic deformation between planning and delivery. Creation of a geometrically resolved view between the endorectal coil MRI and nonendorectal coil image allows information (i.e., prostate gland defined on MRI) to be accurately propagated onto the planning CT. Both rigid and deformable registrations between MRI and nonendorectal CT are challenged by lower soft-tissue contrast resolution of the prostate boundary on the planning CT, limiting its use to guide the transformation. Hensel et al. (2007) developed a technique to enable biomechanical registration of these images for radiotherapy planning, which does not rely on the CT-defined prostate boundary. Multiorgan FEM models were created from images with and without the endorectal coil and consisted of the pelvic bones, bladder, rectum, and prostate gland. The pelvic bones were aligned rigidly and boundary conditions were defined for the bladder and rectum surfaces, allowing the prostate to be deformed according to the biomechanical properties. Registration accuracy of the prostate was within the 2 mm voxel resolution and within the prostate deformation induced by the coil (8.8 mm, range 4.2–13.4 mm). In a similar application, Zhong et al. (2015) applied a hybrid DIR to map prostate contours defined on MR into planning CTs. An intensity-based DIR is applied within a bounding box surrounding the prostate gland in the images. Displacements on the bounding box surface are derived from the intensity based DIR and serve as boundary conditions for the biomechanical FEM model representing anatomy inside the bounding box. The adaptive FEM method preserves the prostate volume and displacements better than the intensity based DIR method without the requirement to delineate any tissues directly on the secondary image (i.e., planning CT). A potential advantage of applying any of these biomechanical techniques is removing the need to define the prostate on CT, which is more prone to contouring uncertainties versus MRI, thereby potentially reducing delineation errors.

5.3.2.2 Consensus Target Definitions

Although the above-mentioned scenarios require targets to be defined for specific patients and treatment plans, the radiation oncology community is also tasked with establishing consensus descriptions of target volumes to ensure that they are uniformly defined across the population. Interpatient image registration can be used to visualize targets defined on individual images in a common geometrically resolved view to appraise the adequacy of guidelines. The extreme range of anatomic variability between patients, however, precludes the use of rigid registration. In addition, commonly utilized metrics to assess concordance of volumes (e.g., overlap measures) do not quantify the spatial location of any discrepancies. Deformable registration, however, can be applied in this setting, inconsistencies in spatial location of nonanatomic material between patients, such as implanted fiducial markers or surgical clips, could potentially introduce nonphysical transformations.

Biomechanical DIR of interpatient images has been applied to validate consensus definitions of the radiotherapy target volume for prostate cancer patients postprostatectomy (Wiltshire et al. 2007). Each patient's surgical prostate bed was delineated on their planning CT based on anatomic boundaries according to a proposed consensus. Using a FEM approach, interpatient deformable registration mapped each patient's target to single reference geometry with boundary constraints applied to the target surfaces. The internal volume, including approximately 25 surgical clips implanted within each prostatectomy area was deformed through the biomechanical model. A single geometrically resolved view of the target for 31 patients was then generated. The distribution of the surgical clips throughout the area defined by anatomic boundaries validated that the proposed consensus accurately represented the desired target, as the surgical bed is considered at high risk of containing microscopic disease. As in this example, it may be advantageous to apply DIR algorithms that are indifferent to the presence of nonanatomic material as their spatial location can vary relative to common anatomic boundaries. Of note, inherent morphological variations between patients are not solely governed by tissue biomechanics and a common assumption of many DIR algorithms (i.e., intensity or biomechanically based) that *the same* tissue is present between images may be violated. In the absence of a ground truth to establish interpatient registration accuracy, biomechanical DIR can offer a transformation that is physically plausible, which may be sufficient in this application.

The use of biomechanical DIR offers a mechanism to integrate imaging from multiple patients to verify the delineation guidelines to represent the intended volume. Such validation is valuable in standardizing care across patient populations, which consequently may reduce uncertainties in clinical outcomes and impact the results of clinical trials.

5.3.2.3 Safety Margin Design

During treatment planning, gross target volumes delineated on the planning CT are geometrically expanded using a *safety* margin. This expansion termed as the PTV is intended to account for the geometric uncertainties present during radiotherapy. Dose is then planned to encompass the PTV to increase the confidence that prescribed doses are delivered to gross and subclinical disease. Geometric uncertainties such as day-to-day motion, or interfraction motion, are not known on a patient-specific basis at the time of treatment planning but they can be estimated from previously treated patients. This knowledge can be gained by performing image registration between the planning CT and daily volumetric images used for image guidance, such as cone-beam CT (CBCT). As rigid registration is routinely used to measure and correct patient positioning and target tumors, residual uncertainties are often due to soft tissue deformations. Therefore, DIR can be used as a tool to quantify deformation and provide detailed population statistics of interfraction motion and deformation of targets that can then be used to design appropriate PTV margins.

Biomechanical DIR has been applied retrospectively to quantify target deformation between treatment-planning images and images acquired during the course of radiotherapy in the prostate, rectum, and liver (Brierley et al. 2011; Nichol et al. 2007; Velec et al. 2014). This application is particularly useful for quantifying the residual errors of targets that are poorly visualized on CBCT, such as liver tumors that are indistinguishable from the surrounding normal liver parenchyma on CBCT. Using biomechanical models of the liver volume, the tumor's position can be predicted without the need to define the tumor on CBCT.

For sites such as the lungs and liver, it is often required to additionally incorporate breathing motion, which can exceed 20 mm in magnitude (Langen and Jones 2001) into PTVs. Intrafraction motion due to patient-specific breathing motion can be quantified during treatment planning using imaging (Velec et al. 2011, 2014). Respiratory correlated 4D CT (4D CT) provides a series of 3D datasets capturing breathing motion; however, as on CBCT, liver tumors are not well visualized without synchronized intravenous contrast. Therefore, 4D quantification of tumor motion is challenging and often surrogates such as the liver or diaphragm are relied on in combination with rigid-body models.

DIR may be used to propagate liver tumor volumes defined on the planning CT images acquired at exhale to the inhale phase of the 4D CT, and the tumor's occupancy volume can form a large component of the PTV. Potential challenges in using DIR for this application include the poor tumor visualization as compared to normal liver on 4D CT without contrast, or when there is variable tumor enhancement (i.e., intravenous contrast is present on the exhale planning CT and not on the inhale 4D CT). Biomechanical liver models are largely insensitive to the tumor-intensity variations on the images and can be used to accurately estimate this 4D motion. A study of 21 patients planned for liver radiotherapy compared tumor motion between the exhale and inhale 4D CT datasets, modeled with both rigid-liver motion and biomechanical DIR (Velec et al. 2011). The FEM liver model containing the tumors was constructed on the exhale-phase planning 4D CT and deformed to the inhale 4D CT with boundary conditions applied to the liver surfaces. In this study, the amplitude of breathing motion estimated using the rigid-body motion of the liver differed by at least 3 mm versus the biomechanical model in more than half of the patients, up to a maximum difference of 10 mm. Six of the seven patients also had multiple intrahepatic tumors with breathing amplitude differences between tumors greater than 3 mm, up to a maximum of 10 mm, due to breathing-induced liver deformation. These data illustrate that PTV margins designed to encompass breathing motion using rigid-body estimates and tumor surrogates may not accurately capture the full extent of motion and deformation. The increased uncertainty may need to be considered and accounted for in further PTV expansions causing greater normal tissue irradiation.

5.3.2.4 Functional Image Assessment

In addition to anatomical images, radiotherapy treatment planning increasingly incorporates metabolic and functional information (e.g., PET, functional MRI) to aid in target delineation or to avoid the highest-functioning normal tissue regions. In some clinical scenarios, it is possible to derive physiological information directly from morphological imaging. Ventilation imaging, for example, mapping the spatial air volume exchange in lung tissues, can be acquired with PET or estimated directly with DIR of the 4D CT images acquired for treatment planning. Registration-based methods typically compute the voxel-wise functionality by the volumetric expansion or intensity variations between exhale and inhale images. Zhong et al. argued that because ventilation is related to lung mechanics, determining the regional compliance, defined as the ventilation per unit stress, will aid in understanding the lung function (Zhong et al. 2011). Images of regional compliance were computed by first applying a Demons intensity based DIR on lung 4D CT images to determine the displacement and volumetric variation of each element, followed by a FEM-based analysis to compute the stress images. Therefore in this application, the biomechanical model is used for 3D stress computation and not the DVF, providing functional information without the need for an additional imaging.

5.3.3 Image-Guided Treatment Delivery

Daily radiotherapy image guidance often relies on 3D soft tissue imaging such as CBCT. These images are rigidly registered to the planning CT to improve targeting through rigid-body corrections. Although DIR can quantify the magnitude of deformation during treatment, correcting the patient's position is limited by the DOF of the treatment couch, which typically permits only translations and rotations. Thus, the role of DIR for online image guidance may be limited. However, where tumors are not visible on CBCT, such as in the liver, DIR can model the tumor position and a local rigid translation can be performed to approximately correct tumor displacements. Biomechanical DIR can predict the impact of interfraction liver deformation on tumor displacement when they are not directly visualized on CBCT. This was studied in 12 patients receiving stereotactic radiotherapy over six fractions with CBCT-based image guidance and rigid registration of the liver (Brock et al. 2008a). Biomechanical model-based DIR retrospectively mapped the tumor position from the planning CT to each CBCT where the actual position was not visible during treatment. Residual differences in the center of mass of the tumors up to 10 mm were observed relative to the planned position (Brock et al. 2008a). One-third of the patients had at least one of the six treatment fractions with a residual tumor displacement exceeding 3 mm, the tolerance level for targeting the liver. These errors were most pronounced in patients with multiple lesions in different lobes of the liver where it was difficult to accurately align all regions of the liver using rigid registration.

Applying biomechanical DIR online would potentially allow for the detection and correction of residual liver tumor displacements on CBCT. Although single-organ DIR is efficient with computation times <1 minute reported (Brock et al. 2008a), the manual segmentation of the liver on CBCT is currently rate limiting in the absence of robust autosegmentation algorithms. To address this limitation, Nguyen et al. (2009) developed a technique to harness the advantages of biomechanical modeling to predict liver tumor motion without the dependence of manual liver segmentation. *A priori* population FEM models of respiratory motion were adapted to patient-specific 4D CT images using narrow region-of-interest windows, termed navigator channels. These channels detect 1D shifts in the image-intensity profiles along the liver boundaries between exhale and inhale images and are used to refine the average population model to generate fully patient-specific 4D models. The navigator channel technique had an equivalent accuracy compared to biomechanical DIR of the patient's 4D imaging yet it was completed within 1–2 min without the need for liver segmentation. These promising results suggest that it could reasonably be applied for online image guidance.

5.3.4 Deformation and Accumulation of Dose

A common limitation of current commercially available treatment planning systems is that dose is calculated on a single CT offering only a static representation of the patient. This is at odds with the fact that radiotherapy is delivered over multiple fractions and for several anatomic sites also under the condition of constant breathing motion. Incorporating intra- and inter-fraction motion and deformation into dose calculations would more accurately reflect the doses delivered to patients.

5.3.4.1 Incorporation of Breathing Motion into Planned Doses

PTVs accounting for breathing motion on static planning CT images are a robust and simple measure to ensure that the tumor receives the prescribed dose under breathing conditions. However, static planned doses that are calculated on a single-planning CT image do not accurately reflect the doses to the surrounding normal tissues that move into and out of the irradiated region. As DIR can be used to model respiratory motion from 4D CT, this motion and deformation can be incorporated into dose calculations. The technique often involves recalculating dose distributions initially optimized on the planning CT (e.g., exhale CT) onto additional 4D CT datasets representing the range of motion between the exhale and inhale. The locations of all the voxels or elements in the case of FEM models are tracked across the images and dose matrices using the DVF. Doses are then summed from all positions with weighting to account for the time spent at each position to determine the accumulated dose over the breathing cycle (Brock et al. 2003). Biomechanical DIR has been well studied on thoracic and abdominal 4D CT data. The ability to accurately perform multiorgan registration, including low-contrast regions (e.g., liver tumors) and modeling complex interfaces (e.g., sliding of the lungs and chest wall), makes it a promising candidate for DIR-based dose accumulation on 4D CT images.

A retrospective analysis was conducted on the actual treatment plans for 21 patients receiving stereotactic liver radiotherapy where multiorgan biomechanical DIR and dose accumulation was performed on 4D CT (Velec et al. 2011). More than half of the patients studied (57%) had dosimetric changes greater than 5% to tumor or normal tissues when comparing the dose distribution accounting for breathing motion to the static dose used clinically. These changes ranged from −14% to 7% for the minimum tumor doses and from −25% to 13% for the maximum normal tissue doses. Knowledge of these deviations could potentially allow clinicians to better optimize the plan and spare more normal tissue, or allow for higher doses to be prescribed to tumors for an equivalent risk of toxicity. Dose distributions incorporating breathing motion were also compared between multiorgan DIR and rigid-body motion of the liver, a more simpler readily accessible technique; however, differences up to 8% in the minimum tumor dose and 7% in the maximum normal tissue doses were observed. This highlights the limitation in using rigid-body motion to accumulate dose for all the relevant abdominal organs.

5.3.4.2 Accurately Accounting for Delivered Doses

The use of 3D imaging for daily image guidance allows the extension of the dose accumulation capability to encompass the interfraction motion. Images acquired just prior to treatment delivery are more representative of the patient versus pretreatment planning images that allow for improved estimates of the doses actually delivered. Applying FEM-based models to accumulate delivered dose to deforming organs was first demonstrated by Yan et al. (1999). For one prostate radiotherapy patient, a FEM model of the rectal wall based on the planning CT and deformed to 14 repeat CT scans acquired twice weekly during the course of treatment. Variations in the doses delivered to individual elements, assumed to represent a functional subunit of the rectal wall, caused variations in the risk of radiation-related toxicity ranging from 0.3% to 94.4%. Such large uncertainties in radiobiological response may explain part of the large variability in the observed treatment outcomes.

Reconstructing delivered doses using biomechanical models is suitable when DIR must be performed on CBCT now routinely used for image guidance where soft tissue contrast is poorer as compared to diagnostic-quality CT images. An additional concern in applying any DIR between planning CT and CBCT is that the later often has a smaller imaging field-of-view, which truncates relevant anatomy. The discrepancy in tissues depicted in the images presents a challenge for image registration for structures that are near or partially extending outside the smaller field-of-view. Biomechanical DIR that relies on an initial FEM-based model derived from the planning CT inherently offers a method to *fill in* the missing tissue on CBCT. As shown in Figure 5.5, the missing tissue can be estimated assuming a local rigid registration of the structure of interest, whereas the anatomy within and extending outside the field-of-view is permitted to deform through the biomechanical properties. This DIR approach was applied in a retrospective study quantifying the delivered doses to liver patients treated with stereotactic radiotherapy (Velec et al. 2012). Multiorgan FEM-based biomechanical DIR was used to register CT and CBCT with boundary conditions applied to contours of the liver, spleen, and body, organs that are easily visualized and predominantly within CBCT. Other nonvisible organs of interest and tumors were deformed through the biomechanical model. The delivered doses, accumulated over the six treatment fractions, accounted for interfraction positing errors, organ deformation, and daily variations in breathing motion. The majority of patients (70%) had substantial changes between the delivered dose and the static planned dose distribution, ranging from −42% to 8% in normal tissues. Such deviations from the planned doses are greater in magnitude and frequency than deviations between planned doses that incorporate 4D CT breathing motion and planned static doses. Knowledge of the delivered doses may be clinically relevant when interpreting tumor response or radiation-induced normal tissue toxicities.

Experiments comparing dose distributions deformed using various intensity-based DIR algorithms have revealed highly variable results on deformable dosimeter phantoms ($\gamma_{3\%/3\,mm}$ range: 53.2%–98.8%) (Yeo et al. 2012). Inaccuracies with intensity based

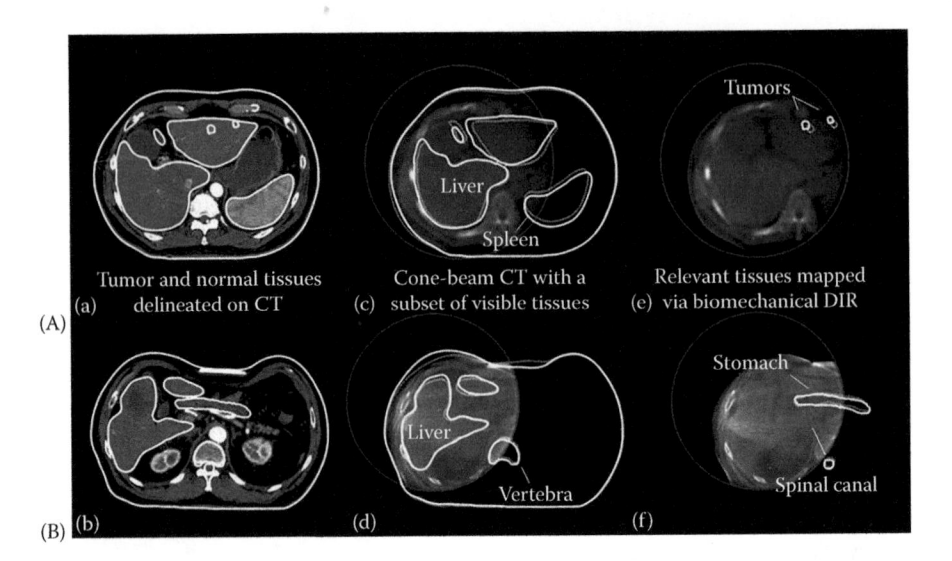

FIGURE 5.5 Two examples ([A] and [B]) of deformable registration between the planning CT ([a] and [b]), on which the patient FEM model is based, and limited field-of-view cone-beam CT ([c] and [d]) where only a subset of organs are visible. Tissues near or extending outside the field-of-view border are accurately mapped through the biomechanical model despite the missing tissue in the image ([e] and [f]). Solid contours define regions on the CT and dashed contours define regions on the CBCT.

DIR can be related to the homogenous internal appearance of the phantoms on imaging (Juang et al. 2013). This is also a challenge for patients when reconstructing doses in tissues with homogenous appearances on images used for guidance (e.g., CBCT). Even with diagnostic-quality imaging, the choice of DIR algorithms for dose reconstruction is crucial. Jamema et al. (2015) compared dose reconstruction on MR images for cervix patients receiving brachytherapy using an optical flow intensity based DIR and a biomechanical DIR driven by the minimization of energy on elastic surface meshes for the bladder and rectum. The differences between the delivered reconstructed doses and planned doses were on average similar in magnitude to differences between doses reconstructed using either DIR algorithm. However, implausible deformations were observed with the intensity based DIR. The application of biomechanical algorithms that are less dependent on the image intensity can be useful in scenarios where intensity-based DIR has not been validated for specific modalities (e.g., MR) or images with poor soft-tissue contrast (e.g., CBCT).

5.3.4.3 Adapting to Patient-Specific Variations in Delivered Doses

Detecting deviations in delivered doses early during the treatment course could enable plan modification through the refinement of margins to account for patient-specific variations and enable reoptimization of the dose distribution. This systematic feedback loop forms a strategy termed as adaptive radiation therapy (Yan et al. 1997). Adaptive

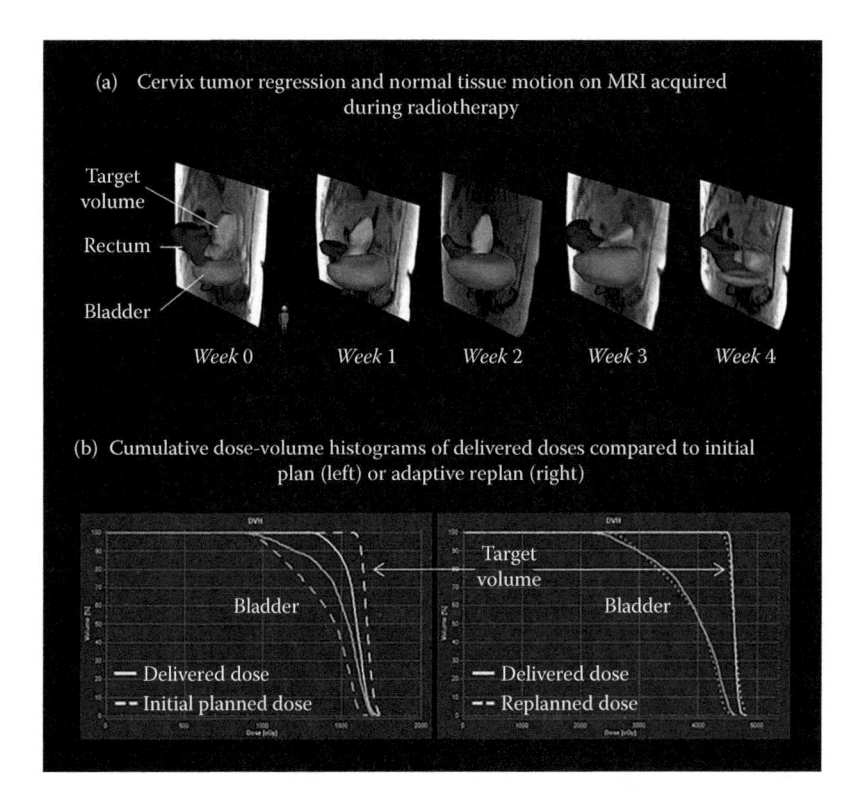

FIGURE 5.6 Impact of organ motion (a) on delivered radiotherapy dose (b). Substantial motion beginning at week 1 causes target underdosage and bladder overdosage, which is corrected with an adaptive replan for the remainder of the treatment.

strategies have been studied extensively using biomechanical DIR to model interfraction accumulated delivered doses in cervix cancer radiotherapy (Figure 5.6) (Lim et al. 2009, 2014; Stewart et al. 2010). Weekly MRIs of the pelvis were used to model motion observed over the five-week course of treatment (Lim et al. 2009). Cervix cancers are subject to daily variations in organ filling and tumors that can regress over several weeks of therapy. For 30 patients who received whole pelvis irradiation, cervix tumor regression ranging from 48% to 96% has been observed between planning and the MR acquired during the last week of treatment. The biomechanical model-based DIR used in this study applied boundary conditions to the normal tissues surfaces and the tumor. Although tumor regression is likely indicative of tissue *changes* rather than motion governed strictly by biomechanics, applying boundary condition to tumor surfaces is performed to account for decreasing tumor volume and improve accuracy of the multiorgan model. The estimated delivered dose to the tumor can then be used to inform whether an adaptive intervention, such as replanning, is required.

Accumulating doses to tumors that exhibit large response is however controversial, as accounting for mass changes using DIR violates the principle of the conservation of energy (Zhong and Chetty 2016). Although the geometric accuracy of the registration appears correct visually, directly applying the underlying DVF to deform doses can lead to an over or underestimation in delivered doses. To minimize this uncertainty, algorithms with local rigidity constraints around tumors to preserve their volumes have been proposed (Zhong and Chetty 2016). Biomechanical model-based DIR can also be used without explicit mapping of the tumor regression, and instead rely on the deformation induced by the surrounding normal tissues to perform image registration. Further research that characterizes tumor dynamics, for example, whether tumors dissolve or elastically contract may help guide the selection of appropriate DIR technique that reduces uncertainties in dose accumulation.

5.3.5 Treatment-Response Analysis

5.3.5.1 Classifying Tumor Response to Radiation therapy

Evaluating tumor responses to radiotherapy on follow-up diagnostic imaging is challenging as tissues can deform due to physiologic motion and positioning in addition to radiation-induced changes. Following stereotactic radiotherapy of lung tumors, for example, normal lung parenchyma can become fibrotic, which has the dual effect of deforming the anatomy and obscuring the tumor. For liver tumors, normal liver subvolumes that received high-radiation doses can become fibrotic, whereas untreated volumes can hypertrophy causing deformation. Contrast-enhanced follow-up imaging of the liver has also demonstrated that radiation reactions in the normal liver undergo temporal variations appearance, ranging from hypo- to hyperintense (Herfarth et al. 2003). Combined, these effects can decrease the confidence with which clinicians can evaluate tumor response or failure to treatment.

DIR between pretreatment or planning imaging and follow-up imaging can improve registration accuracy by accounting for radiation-induced deformations. The geometrically resolved views can facilitate image interpretation, whereas the DVF can augment standard assessment methods used in clinic (e.g., bidimensional tumor measurements) by inherently providing 3D measurement of variations in spatial response. Biomechanical model-based DIR can resolve the deformation of the tumor-bearing organ without the need to consider the substantial patient-specific variations in image intensity on follow-up imaging.

The feasibility of registering planning CT images and diagnostic CT images acquired 3 months followed by radiation therapy was demonstrated on 5 liver cancer patients (Brock et al. 2006). Biomechanical DIR-generated geometrically resolved views of the follow-up images based on the liver surface contours to account for radiation-induced deformation. A second registration was then performed based on tumor surfaces alone to permit analysis of the spatial variation in response. Compared to the planning CT, three patients had residual tumor surfaces on the geometrically resolved follow-up CT where 13%–89% of the surface area had vector displacements >5 mm and up to 77% of the surfaces had displacements >10 mm. Although such detailed analyses are not routinely performed in the clinical setting, they could potentially be correlated with the heterogeneous delivered doses to better assess spatial responses.

Radiation-induced soft tissue changes observed during the course of treatment can limit the accuracy of biomechanical models, therefore providing a rationale to move away from strictly image-based response to one that incorporates radiobiological response. Al-Mayah developed a framework to include planned radiotherapy doses into FEM models (Al-Mayah et al. 2015). Five head and neck patients with pre- and postradiotherapy MRI exhibiting mean volume changes of 34% and 51% to tumors and salivary glands, respectively, were studied. Each patient's multiorgan FEM was driven by the rigid displacement of bony anatomy and deformation of the external surface contours. Additional loads were applied to tumor and salivary glands to simulate dose-induced shrinkage, based on the regions' volume and planned doses to the nodes comprising each region. The loads were given by the general equation $\Delta V = V\alpha(D)D$ where ΔV is the volume change, $\alpha(D)$ is the coefficient of dose volume change, and D is the planned dose. The average dose-response coefficient values ranged from -0.0143 to -0.184 cm^{3}Gy^{-1}. On the basis of an analogous engineering concept, the dose effect is not applied as a displacement but as a thermal load without spatial constraint. The average DSC of the four salivary glands increased from 0.53, 0.55, 0.32, and 0.37 with the standard FEM approach to 0.68, 0.68, 0.51, and 0.49 with the additional dose effect. This technique has demonstrated the extensibility of biomechanical model to incorporate dose responses. Reducing registration errors would also allow for accurate integration of data from the functional imaging, such as diffusion-weighted MRI, to be incorporated into dose-response models.

5.3.5.2 Correlating Radiation Doses with Clinical Outcomes

The majority of clinical evidence describing outcomes following radiotherapy in the literature is inherently based on the assumption that delivered doses are identical to the planned dose distribution. Deviations in delivered doses previously described potentially confound the ability to accurately predict clinical outcomes, understand dose-volume relationships, and develop radiobiological models. DIR-based delivered dose reconstruction has therefore been hypothesized to better predict outcomes than planned doses. This was investigated by Swaminath et al. (2015) who correlated delivered doses, reconstruction using biomechanical DIR, with local control in liver metastases patients treated with stereotactic radiotherapy. Delivered tumor dose was a significantly *stronger predictor* for local control than planned dose in all statistical and goodness-of-fit models tested. This work provides evidence that decreasing the uncertainty in dosimetry aids in the interpretation of clinical outcomes. Although this application of DIR is not unique to biomechanical modeling, the application of biomechanical DIR was shown throughout the chapter to have great utility in reducing uncertainties in liver radiotherapy. There is a strong interest in radiation oncology to expand the role of dose reconstruction techniques to develop dose-response models for normal tissue toxicity (Jaffray et al. 2010).

In some clinical scenarios, even planned 3D dosimetry is not available to investigate dose responses due to patients historically being treated prior to the development and adoption of 3D imaging and 3D radiotherapy treatment planning. Model-based approaches have been developed to estimate the 3D geometry of historic patients based on their 2D imaging record. Such is the case for Hodgkin's lymphoma patients for whom

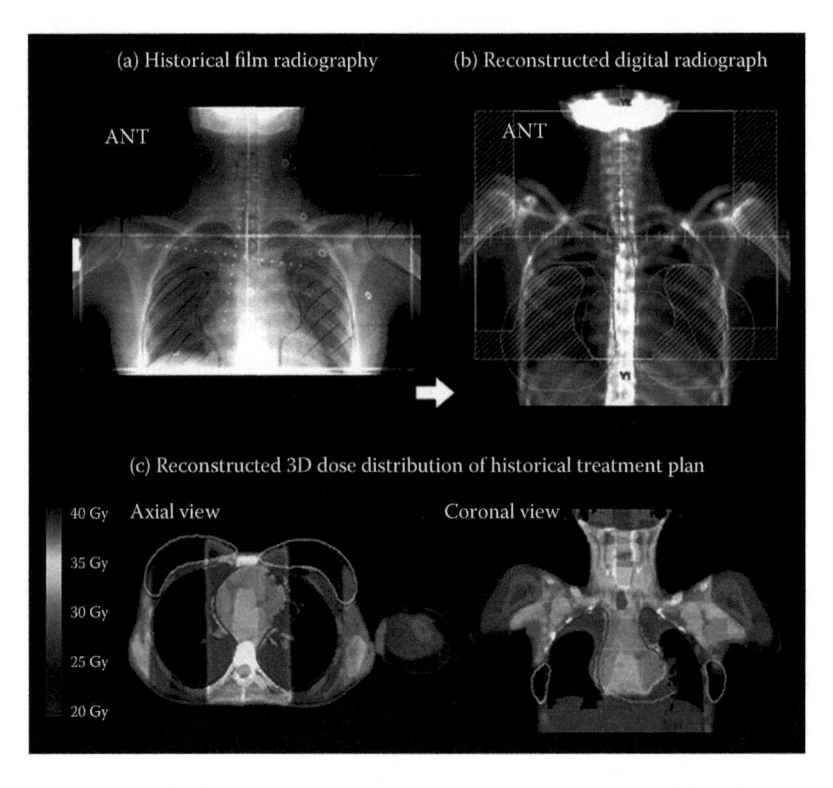

FIGURE 5.7 Result of CT reconstruction using biomechanical FEM models and a navigator channel technique. For a historic patient with 2D planning films (a), a CT dataset is modeled enabling the recreation of the treatment beams (b) and calculation of a 3D dose distribution (c) from which normal tissue doses can be estimated. (Adapted from Ng, A. et al., *Int. J. Radiat. Oncol. Biol. Phys.*, 84, s557–s563, 2012. With permission.)

secondary cardiac morbidity and cancers of the lungs and breasts can occur decades after irradiation (Gustavsson et al. 2003). For these patients lack planning CT images and 3D dose distributions, Ng et al. (2010) developed population-based 3D biomechanical FEM models, which were deformed into patient-specific 3D models using a navigator channel technique and extracted anatomic features from their historic 2D planning films. This allowed for the reconstruction of individualized planning CT images and calculation of 3D organ dosimetry (Figure 5.7). The technique accurately reconstructed known lung and heart volumes with mean DSC of 0.89 (Ng et al. 2010, 2012b). This translated into median reconstructed 3D dosimetry errors of <2% for the relevant dose metrics to the lung, heart, and breast tissues, which are similar in magnitude to other sources of uncertainty (e.g., breathing motion) (Ng et al. 2012a). The 3D dosimetry for 10 historical patients treated using a traditional 2D *full-mantle mediastinum* technique was demonstrated to result in planned normal tissue doses nearly double that of contemporary patients treated with smaller *involved-field* 3D treatment plans (Ng et al. 2012a).

As planned dosimetry metrics do not incorporate motion into their calculation, they do not require the level of DIR accuracy required for dose mapping. The dosimetry is dependent on accurate definition of the organ boundaries and that the CT intensities are mapped in a physically plausible manner for dose calculation. Although it is not possible to quantify internal organ accuracy on the historic patients due to lack of a 3D imaging, this application can be expected to be less sensitive to DIR errors than dose-accumulation applications because the CT intensities within individual organs are relatively homogenous. This framework demonstrated the capability of biomechanical DIR to build predictive dose-toxicity models and include patients lacking 3D radiotherapy dosimetry but for whom mature outcome data are available.

5.3.6 Correlating Imaging to Pathological Specimens

One of the largest uncertainties in radiation oncology is caused by a lack of knowledge of precisely where microscopic disease lies on planning imaging and how to interpret the imaging signal to identify the tumor boundary. Methods to overcome this include fusing tumor representations from complementary image modalities, *overcontouring* at ambiguous tumor borders, and expanding PTV margins. Validation of image-based diagnosis and response assessment could potentially be performed by correlating clinical images to the pathological specimens considered as the *gold standard*. Multiple sources of deformation occur between clinical imaging and pathology making DIR a necessity.

In the prostatectomy setting, for example, these include tissue resections, microtome sectioning, sample shrinkage from fixation, and dehydration processes and blood loss and pressure of the endorectal coil. Biomechanical models have been proposed to resolve deformation while preserving the spatial integrity of information without the need to establish a correspondence between the substantially different imaging modalities (i.e., *in vivo* MRI and digital histology slides). For patients who underwent radical prostatectomies, Samavati et al. (2011) developed a multistep process to align histological slides to *in vivo* MR images acquired with an endorectal coil 3–5 weeks prior to surgery. A 3D tetrahedral FEM model of the prostate is constructed from histology imaging with homogenous linear elastic material properties assigned initially. This model incrementally deforms the histology model to *ex vivo* MR of the fixed prostate specimen, then to the *ex vivo* MR of the fresh prostate specimen, and finally to the *in vivo* MR. In this implementation, heterogeneous material properties were incorporated into the process based on MR elastography of each specimen (McGrath et al. 2012). Between histology and *in vivo* MR, the proposed process achieved an average TRE of 2.3 mm within the *in vivo* MR slice thickness with a particular improvement in DIR over rigid registration when intraprostatic lesions were adjacent to the prostate surfaces.

5.4 Summary

In summary, biomechanical-based DIR algorithms have been developed to reduce uncertainties in radiotherapy, improve the design of treatment planning and delivery, offer new adaptive strategies, and enhance understanding and modeling of treatment response. On patient data, the accuracy of biomechanical FEM models is on the order of the image resolution that suggests improved accuracy compared to currently available techniques. Further evaluation is required to rigorously validate accuracy of these models, in particular for the more complex tasks such as dose accumulation to ensure that additional uncertainties are not introduced into treatment processes.

References

Al-Mayah, A., J. Moseley, S. Hunter, and K. Brock. 2015. Radiation dose response simulation for biomechanical-based deformable image registration of head and neck cancer treatment. *Phys Med Biol* 60 (21):8481–8489.

Al-Mayah, A., J. Moseley, S. Hunter, M. Velec, L. Chau, S. Breen, and K. Brock. 2010. Biomechanical-based image registration for head and neck radiation treatment. *Phys Med Biol* 55 (21):6491–6500.

Al-Mayah, A., J. Moseley, M. Velec, and K. Brock. 2011. Toward efficient biomechanical-based deformable image registration of lungs for image-guided radiotherapy. *Phys Med Biol* 56 (15):4701–4713.

Al-Mayah, A., J. Moseley, M. Velec, and K. Brock. 2009a. Deformable modeling of human liver with contact surface. In *Science and Technology for Humanity (TIC-STH), 2009 IEEE Toronto International Conference*. Toronto, ON, IEEE.

Al-Mayah, A., J. Moseley, M. Velec, and K. Brock. 2009b. Sliding characteristic and material compressibility of human lung: Parametric study and verification. *Med Phys* 36 (10):4625–4633.

Alterovitz, R., K. Goldberg, J. Pouliot, I. C. Hsu, Y. Kim, S. M. Noworolski, and J. Kurhanewicz. 2006. Registration of MR prostate images with biomechanical modeling and nonlinear parameter estimation. *Med Phys* 33 (2):446–454. doi:10.1118/1.2163391.

Bharatha, A., M. Hirose, N. Hata, S. K. Warfield, M. Ferrant, K. H. Zou, E. Suarez-Santana et al. 2001. Evaluation of three-dimensional finite element-based deformable registration of pre- and intraoperative prostate imaging. *Med Phys* 28 (12):2551–2560. doi:10.1118/1.1414009.

Boubaker, M. B., M. Haboussi, J. F. Ganghoffer, and P. Aletti. 2009. Finite element simulation of interactions between pelvic organs: Predictive model of the prostate motion in the context of radiotherapy. *J Biomech* 42 (12):1862–1868.

Brierley, J. D., L. A. Dawson, E. Sampson, A. Bayley, S. Scott, J. L. Moseley, T. Craig et al. 2011. Rectal motion in patients receiving preoperative radiotherapy for carcinoma of the rectum. *Int J Radiat Oncol Biol Phys* 80 (1):97–102.

Brock, K. K. 2010. Results of a multi-institution deformable registration accuracy study (MIDRAS). *Int J Radiat Oncol Biol Phys* 76 (2):583–596.

Brock, K. K., L. A. Dawson, M. B. Sharpe, D. J. Moseley, and D. A. Jaffray. 2006. Feasibility of a novel deformable image registration technique to facilitate classification, targeting, and monitoring of tumor and normal tissue. *Int J Radiat Oncol Biol Phys* 64 (4):1245–1254.

Brock, K. K., M. Hawkins, C. Eccles, J. L. Moseley, D. J. Moseley, D. A. Jaffray, and L. A. Dawson. 2008. Improving image-guided target localization through deformable registration. *Acta Oncol* 47 (7):1279–1285.

Brock, K. K., D. L. McShan, R. K. Ten Haken, S. J. Hollister, L. A. Dawson, and J. M. Balter. 2003. Inclusion of organ deformation in dose calculations. *Med Phys* 30 (3):290–295.

Brock, K. K., J. Moseley, D. Moseley, and D. Jaffray. 2005. 278 A ray-driven approach to generate a deformed image for image-guided radiotherapy. *Radiother Oncol* 76 (S2):S129–S130.

Brock, K. K., A. M. Nichol, C. Menard, J. L. Moseley, P. R. Warde, C. N. Catton, and D. A. Jaffray. 2008. Accuracy and sensitivity of finite element model-based deformable registration of the prostate. *Med Phys* 35 (9):4019–4025.

Brock, K. K., M. B. Sharpe, L. A. Dawson, S. M. Kim, and D. A. Jaffray. 2005. Accuracy of finite element model-based multi-organ deformable image registration. *Med Phys* 32 (6):1647–1659.

Cazoulat, G., D. Owen, M. M. Matuszak, J. M. Balter, and K. K. Brock. 2016. Biomechanical deformable image registration of longitudinal lung CT images using vessel information. *Phys Med Biol* 61 (13):4826–4839.

Chai, X., M. van Herk, M. C. Hulshof, and A. Bel. 2012. A voxel-based finite element model for the prediction of bladder deformation. *Med Phys* 39 (1):55–65.

Chi, Y., J. Liang, and D. Yan. 2006. A material sensitivity study on the accuracy of deformable organ registration using linear biomechanical models. *Med Phys* 33 (2):421–433.

Crouch, J. R., S. M. Pizer, E. L. Chaney, Y. C. Hu, G. S. Mageras, and M. Zaider. 2007. Automated finite-element analysis for deformable registration of prostate images. *IEEE Trans Med Imaging* 26 (10):1379–1390.

Gustavsson, A., B. Osterman, and E. Cavallin-Stahl. 2003. A systematic overview of radiation therapy effects in Hodgkin's lymphoma. *Acta Oncol* 42 (5–6):589–604.

Hensel, J. M., C. Menard, P. W. Chung, M. F. Milosevic, A. Kirilova, J. L. Moseley, M. A. Haider, and K. K. Brock. 2007. Development of multiorgan finite element-based prostate deformation model enabling registration of endorectal coil magnetic resonance imaging for radiotherapy planning. *Int J Radiat Oncol Biol Phys* 68 (5):1522–1528.

Herfarth, K. K., H. Hof, M. L. Bahner, F. Lohr, A. Hoss, G. van Kaick, M. Wannenmacher, and J. Debus. 2003. Assessment of focal liver reaction by multiphasic CT after stereotactic single-dose radiotherapy of liver tumors. *Int J Radiat Oncol Biol Phys* 57 (2):444–451.

Jaffray, D. A., P. E. Lindsay, K. K. Brock, J. O. Deasy, and W. A. Tome. 2010. Accurate accumulation of dose for improved understanding of radiation effects in normal tissue. *Int J Radiat Oncol Biol Phys* 76 (3 Suppl):S135–S139.

Jamema, S. V., U. Mahantshetty, E. Andersen, K. O. Noe, T. S. Sorensen, J. F. Kallehauge, S. K. Shrivastava, D. D. Deshpande, and K. Tanderup. 2015. Uncertainties of deformable image registration for dose accumulation of high-dose regions in bladder and rectum in locally advanced cervical cancer. *Brachytherapy* 14 (6):953–962. doi:10.1016/j.brachy.2015.08.011.

Juang, T., S. Das, J. Adamovics, R. Benning, and M. Oldham. 2013. On the need for comprehensive validation of deformable image registration, investigated with a novel 3-dimensional deformable dosimeter. *Int J Radiat Oncol Biol Phys* 87 (2):414–421.

Langen, K. M., and D. T. Jones. 2001. Organ motion and its management. *Int J Radiat Oncol Biol Phys* 50 (1):265–278. doi:10.1016/S0360-3016(01)01453-5 [pii].

Li, M., E. Castillo, X. L. Zheng, H. Y. Luo, R. Castillo, Y. Wu, and T. Guerrero. 2013. Modeling lung deformation: A combined deformable image registration method with spatially varying Young's modulus estimates. *Med Phys* 40 (8):081902.

Li, M., K. Miller, G. R. Joldes, R. Kikinis, and A. Wittek. 2016. Biomechanical model for computing deformations for whole-body image registration: A meshless approach. *Int J Numer Method Biomed Eng* 32(12). doi:10.1002/cnm.2771.

Li, S., C. Glide-Hurst, M. Lu, J. Kim, N. Wen, J. N. Adams, J. Gordon, I. J. Chetty, and H. Zhong. 2013. Voxel-based statistical analysis of uncertainties associated with deformable image registration. *Phys Med Biol* 58 (18):6481–6494. doi:10.1088/0031-9155/58/18/6481.

Liang, J., and D. Yan. 2003. Reducing uncertainties in volumetric image based deformable organ registration. *Med Phys* 30 (8):2116–2122. doi:10.1118/1.1587631.

Lim, K., V. Kelly, J. Stewart, J. Xie, Y. B. Cho, J. Moseley, K. Brock, A. Fyles, A. Lundin, H. Rehbinder, and M. Milosevic. 2009. Pelvic radiotherapy for cancer of the cervix: Is what you plan actually what you deliver? *Int J Radiat Oncol Biol Phys* 74 (1):304–312.

Lim, K., J. Stewart, V. Kelly, J. Xie, K. K. Brock, J. Moseley, Y. B. Cho et al. 2014. Dosimetrically triggered adaptive intensity modulated radiation therapy for cervical cancer. *Int J Radiat Oncol Biol Phys* 90 (1):147–154.

McGrath, D. M., W. D. Foltz, A. Al-Mayah, C. J. Niu, and K. K. Brock. 2012. Quasi-static magnetic resonance elastography at 7 T to measure the effect of pathology before and after fixation on tissue biomechanical properties. *Magn Reson Med* 68 (1):152–165.

Menard, C., D. Iupati, J. Publicover, J. Lee, J. Abed, G. O'Leary, A. Simeonov et al. 2015. MR-guided prostate biopsy for planning of focal salvage after radiation therapy. *Radiology* 274 (1):181–191.

Neylon, J., X. Qi, K. Sheng, R. Staton, J. Pukala, R. Manon, D. A. Low, P. Kupelian, and A. Santhanam. 2015. A GPU based high-resolution multilevel biomechanical head and neck model for validating deformable image registration. *Med Phys* 42 (1):232–43.

Ng, A., K. K. Brock, M. B. Sharpe, J. L. Moseley, T. Craig, and D. C. Hodgson. 2012a. Individualized 3D reconstruction of normal tissue dose for patients with long-term follow-up: A step toward understanding dose risk for late toxicity. *Int J Radiat Oncol Biol Phys* 84 (4):e557–e563.

Ng, A., T. N. Nguyen, J. L. Moseley, D. C. Hodgson, M. B. Sharpe, and K. K. Brock. 2010. Reconstruction of 3D lung models from 2D planning data sets for Hodgkin's lymphoma patients using combined deformable image registration and navigator channels. *Med Phys* 37 (3):1017–1028.

Ng, A., T. N. Nguyen, J. L. Moseley, D. C. Hodgson, M. B. Sharpe, and K. K. Brock. 2012b. Navigator channel adaptation to reconstruct three dimensional heart volumes from two dimensional radiotherapy planning data. *BMC Med Phys* 12:1.

Nguyen, T. N., J. L. Moseley, L. A. Dawson, D. A. Jaffray, and K. K. Brock. 2009. Adapting liver motion models using a navigator channel technique. *Med Phys* 36 (4):1061–1073.

Nichol, A. M., K. K. Brock, G. A. Lockwood, D. J. Moseley, T. Rosewall, P. R. Warde, C. N. Catton, and D. A. Jaffray. 2007. A magnetic resonance imaging study of prostate deformation relative to implanted gold fiducial markers. *Int J Radiat Oncol Biol Phys* 67 (1):48–56.

Samavati, N., D. M. McGrath, J. Lee, T. van Kwast, M. Jewett, C. Menard, and K. K. Brock. 2011. Biomechanical model-based deformable registration of MRI and histopathology for clinical prostatectomy. *J Pathol Inform* 2:S10. doi:10.4103/2153-3539.92035.

Samavati, N., M. Velec, and K. Brock. 2015. A hybrid biomechanical intensity based deformable image registration of lung 4DCT. *Phys Med Biol* 60 (8):3359–3373.

Stewart, J., K. Lim, V. Kelly, J. Xie, K. K. Brock, J. Moseley, Y. B. Cho et al. 2010. Automated weekly replanning for intensity-modulated radiotherapy of cervix cancer. *Int J Radiat Oncol Biol Phys* 78 (2):350–358.

Swaminath, A., C. Massey, J. D. Brierley, R. Dinniwell, R. Wong, J. J. Kim, M., K. K. Brock, and L. A. Dawson. 2015. Accumulated delivered dose response of stereotactic body radiation therapy for liver metastases. *Int J Radiat Oncol Biol Phys* 93 (3):639–648.

Tehrani, J. N., Y. Yang, R. Werner, W. Lu, D. Low, X. Guo, and J. Wang. 2015. Sensitivity of tumor motion simulation accuracy to lung biomechanical modeling approaches and parameters. *Phys Med Biol* 60 (22):8833–8849.

Velec, M., T. Juang, J. L. Moseley, M. Oldham, and K. K. Brock. 2015. Utility and validation of biomechanical deformable image registration in low-contrast images. *Pract Radiat Oncol* 5 (4):e401–e408.

Velec, M., J. L. Moseley, T. Craig, L. A. Dawson, and K. K. Brock. 2012. Accumulated dose in liver stereotactic body radiotherapy: Positioning, breathing, and deformation effects. *Int J Radiat Oncol Biol Phys* 83 (4):1132–1140.

Velec, M., J. L. Moseley, L. A. Dawson, and K. K. Brock. 2014. Dose escalated liver stereotactic body radiation therapy at the mean respiratory position. *Int J Radiat Oncol Biol Phys* 89 (5):1121–1128.

Velec, M., J. L. Moseley, C. L. Eccles, T. Craig, M. B. Sharpe, L. A. Dawson, and K. K. Brock. 2011. Effect of breathing motion on radiotherapy dose accumulation in the abdomen using deformable registration. *Int J Radiat Oncol Biol Phys* 80 (1):265–272.

Voroney, J. P., K. K. Brock, C. Eccles, M. Haider, and L. A. Dawson. 2006. Prospective comparison of computed tomography and magnetic resonance imaging for liver cancer delineation using deformable image registration. *Int J Radiat Oncol Biol Phys* 66 (3):780–791.

Werner, R., J. Ehrhardt, R. Schmidt, and H. Handels. 2009. Patient-specific finite element modeling of respiratory lung motion using 4D CT image data. *Med Phys* 36 (5):1500–1511.

Wiltshire, K. L., K. K. Brock, M. A. Haider, D. Zwahlen, V. Kong, E. Chan, J. Moseley et al. 2007. Anatomic boundaries of the clinical target volume (prostate bed) after radical prostatectomy. *Int J Radiat Oncol Biol Phys* 69 (4):1090–1099.

Wu, Q., J. Liang, and D. Yan. 2006. Application of dose compensation in image-guided radiotherapy of prostate cancer. *Phys Med Biol* 51 (6):1405–1419. doi:10.1088/0031-9155/51/6/003.

Yan, D., D. A. Jaffray, and J. W. Wong. 1999. A model to accumulate fractionated dose in a deforming organ. *Int J Radiat Oncol Biol Phys* 44 (3):665–675.

Yan, D., F. Vicini, J. Wong, and A. Martinez. 1997. Adaptive radiation therapy. *Phys Med Biol* 42 (1):123–32.

Yeo, U. J., M. L. Taylor, J. R. Supple, R. L. Smith, L. Dunn, T. Kron, and R. D. Franich. 2012. Is it sensible to "deform" dose? 3D experimental validation of dose-warping. *Med Phys* 39 (8):5065–5072. doi:10.1118/1.4736534.

Zhong, H., and I. J. Chetty. 2016. Caution must be exercised when performing deformable dose accumulation for tumors undergoing mass changes during fractionated radiotherapy. *Int J Radiat Oncol Biol Phys* (In press).

Zhong, H., J. Y. Jin, M. Ajlouni, B. Movsas, and I. J. Chetty. 2011. Measurement of regional compliance using 4DCT images for assessment of radiation treatment. *Med Phys* 38 (3):1567–1578. doi:10.1118/1.3555299.

Zhong, H., J. Kim, H. Li, T. Nurushev, B. Movsas, and I. J. Chetty. 2012. A finite element method to correct deformable image registration errors in low-contrast regions. *Phys Med Biol* 57 (11):3499–3515.

Zhong, H., T. Peters, and J. V. Siebers. 2007. FEM-based evaluation of deformable image registration for radiation therapy. *Phys Med Biol* 52 (16):4721–4738. doi:10.1088/0031-9155/52/16/001.

Zhong, H., E. Weiss, and J. V. Siebers. 2008. Assessment of dose reconstruction errors in image-guided radiation therapy. *Phys Med Biol* 53 (3):719–736. doi:10.1088/0031-9155/53/3/013.

Zhong, H., N. Wen, J. J. Gordon, M. A. Elshaikh, B. Movsas, and I. J. Chetty. 2015. An adaptive MR-CT registration method for MRI-guided prostate cancer radiotherapy. *Phys Med Biol* 60 (7):2837–2851. doi:10.1088/0031-9155/60/7/2837.

6

3D Ultrasound-Guided Interventions

Aaron Fenster,
Jessica Rodgers,
Justin Michael, and
Derek Gillies

6.1 Introduction

Interventional procedures in which needles and therapy applicators are introduced to either obtain a biopsy specimen or to deliver ablation therapy (e.g., liver) require real-time guidance of needles to their targets and knowledge of the applicators' tip before ablation therapy can be initiated. Imaging techniques such as X-ray fluoroscopy, magnetic resonance (MR), computed tomography (CT), and ultrasound (US) are used for various applications; however, procedures that require real-time guidance use US imaging because of its ability to provide images at 15–60 Hz. Although this approach is useful, US imaging cannot provide sufficient lesion contrast to the organ or lesion for many applications. Thus, MR and CT are used to produce an image to identify a target, which is registered to the intraoperative US image thereby combining the benefit of real-time US imaging with the high-lesion contrast from MR or CT images. This approach is now used in a variety of interventional applications such as prostate biopsy and focal liver ablation.

All interventional procedures require that the user views the needle as it is being directed to its target. Thus, the biopsy/therapy procedure requires that the user holds the US transducer in contact with the body with one hand, and the needle insertion tool or therapy tool in the other hand. The 2D US image stream is then aligned with the trajectory of the needle so that it is continuously visible in the US image stream as the needle is being advanced into the tissue. Typically, alignment is ensured by clip-on guides on the transducer causing the needle to be inserted inplane of the US image.

Although the needle/tool insertion with 2D US image guidance is used extensively and can be highly effective, it has limitations, which may compromise the interventional procedure. These limitations are because of the use of 2D US intraoperatively. The following are some of the limitations that are being addressed by researchers and companies:

1. Because of the thickness of the US plane (ranging from 1 to 5 mm depending on the distance from the transducer and the elevational resolution of the transducer); the needle may leave the plane of the US image due to needle deflection or misalignment of the plane with the trajectory of the needle. Thus, it is possible to misinterpret the location of the needle/tool tip as the point where the needle leaves the plane of the image. For a biopsy needle with a 22 mm throw, any misinterpretation of the location of the tip could be dangerous to both the patient (needle penetration of the chest in cases of breast biopsy) and the clinician (exit of the needle from the breast and entering the hand of the radiologist). In cases of therapy, the ablation may be delivered to the incorrect location.

2. In therapy application, the tool's location that is relative to the target (tumor) and surrounding sensitive structures must be verified and compared to the therapy plan (e.g., rectal wall and neurovascular bundle in cases of prostate focal ablation, and major blood vessels in cases of focal liver ablation). Thus, a 3D imaging modality must be used to assess the anatomy relative to the tool for the therapy to be safe and effective. Although this can be performed effectively using MR or CT, these add to the cost and complexity of the procedure. Thus, an alternative approach using 3D or 4D US may offer an alternative approach.

3. Some therapy procedures such as thermal or cryoablation require monitoring of the therapy progress over times that can extent to 10 min or longer per ablation. As the margin of the ablation zone must be monitored, a 3D imaging approach provides the best method. CT and MR imaging can be used for the verification of needle/tool location, but 3D US would provide a cost-effective approach.

Although conventional US transducers are being used for guiding needles/tools into tissues for biopsy and therapy, verifying their location that is relative to the target and monitoring the progress of ablation procedures is critical as these procedures are prone to inaccuracy and variability depending on the radiologist's experience. Manual needle identification of the tip and trajectory requires identification and segmentation of the needle. Mental identification and manual segmentation is time-consuming and prone to error and variability as it is carried out under a limited time during the procedure. Thus, many investigators and commercial companies have developed various algorithms to segment the needle or therapy tool (Draper et al. 2000; Ding et al. 2003; Qiu et al. 2013; Hrinivich et al. 2016). These techniques include real-time segmentation of the needle from real-time 2D US images for use in biopsy techniques and segmentation from 3D US images for navigation and verification of the location of the needle relative to the target. Techniques to produce 3D US images have been described extensively in review articles and books (Downey and Fenster 1998; Nelson and Pretorius 1998; Nelson et al. 1999; Fenster and Downey 2000a, Fenster and Downey 2000b, Fenster et al. 2001). Thus, the reader can refer to these for details in these technologies.

6.2 3D Ultrasound-Guided Prostate Biopsy

Prostate cancer (PCa) is the second leading cause of death due to cancer in men worldwide and accounts for between 2.1% and 15.2% of all cancer deaths (Silverberg et al. 1990; Abbas and Scardino 1997; Stewart and Wild, 2014). PCa is curable if diagnosed at an early stage, and can be treated effectively even at later stages. Thus, early diagnosis, accurate staging of PCa, and selecting appropriate therapies are critical to the patient's well-being.

The definitive method for diagnosing PCa involves needle-based prostate biopsy using transrectal US (TRUS) guidance. As PCa is difficult to detect using TRUS imaging, the accepted method is to obtain biopsy samples from predetermined regions of the prostate using approximately 12 cores. Although this method has been used extensively, it is suboptimal, resulting in false negative rates ranging as high as 25% (Rabbani et al. 1998). For this reason, methods have been developed over the past five years to obtain multiparametric MR images of the patient's prostate, identify the suspicious lesions to be biopsied, and fuse the MR image with an intraoperative TRUS image to allow guidance of the biopsy needle to the MR-identified suspicious target (Natarajan et al. 2011). Thus, methods for generating a 3D TRUS image, registration between MR and TRUS, and accurate guidance systems are needed.

We have developed a mechanical 3D US-guided biopsy system that makes use of MR images obtained before the biopsy procedure, which are registered to the intraprocedural 3D TRUS image to allow targeting of MR identified tumors, but guided by US imaging (Bax et al. 2008). Our mechanical guidance system is based on an articulated multijointed stabilizer and a transducer-tracking mechanism that has four degrees-of-freedom (DOF) and has an adaptable cradle to support any commercially available end-firing TRUS transducers used for prostate biopsy. The system allows real-time tracking and recording of the 3D position and orientation of the biopsy needle as the physician manipulates the TRUS transducer (Figure 6.1).

To perform a 3D TRUS-guided prostate biopsy, a conventional end-firing TRUS transducer is mounted onto the tracking assembly. The physician inserts the TRUS transducer into the patient's rectum and rotates the transducer 200 degrees about its

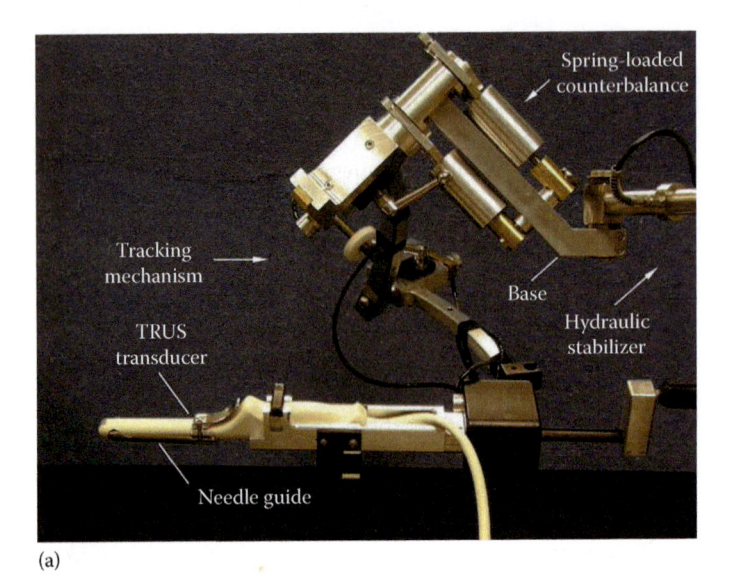

(a)

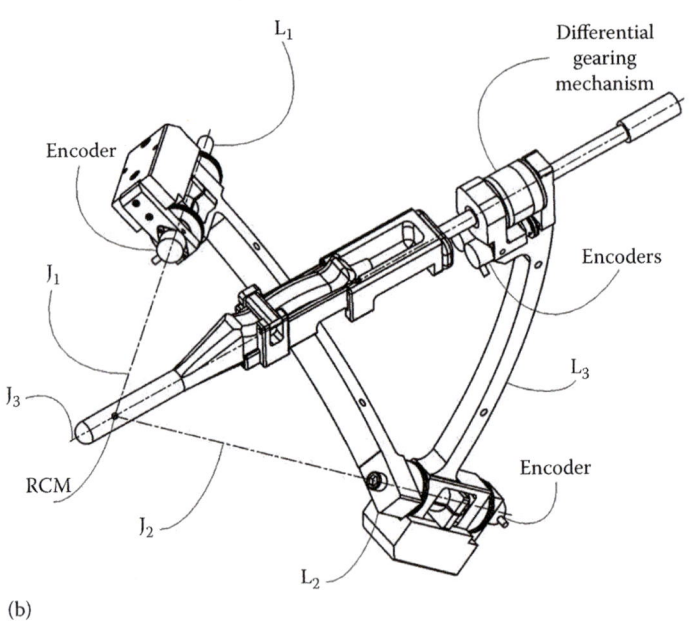

(b)

FIGURE 6.1 (a) 3D TRUS-guided prostate biopsy tracking subsystem with all components labeled. The TRUS transducer is rotated about a remote center-of-motion (RCM), which is at the transducer tip. (b) Schematic diagram of the mechanical-tracking subsystem, which supports the TRUS transducer and the attached cradle. The TRUS probe motion is constrained to three rotational degrees-of-freedom and one degree of translation along the axis of the probe. (From Bax, J. et al., *Med. Phys.*, 35, 5397–5410, 2008.)

longitudinal axis to generate a 3D TRUS image. The 3D TRUS image is then registered to the preprocedural acquired MR image with the outlined tumor to be targeted. After the biopsy targets are selected, the trajectory of the biopsy needle is guided by the system and fired to obtain a core. The location of the biopsy is then recorded automatically.

Our 3D TRUS-based biopsy system is composed of a mechanical tracking subsystem for guidance, 3D TRUS image acquisition and reconstruction, 3D TRUS to MR registration and image fusion, and visual guidance software allowing accurate targeting of the lesion (Bax et al. 2008). The following are details of the system.

6.2.1 Mechanical Tracking Subsystem

The mechanical subsystem with the 3D TRUS scanner allows the user to manipulate the TRUS scanner manually while this subsystem continually tracks the location of the TRUS transducer. This tracking system uses a remote center-of-motion (RCM) mechanism providing four DOF: translation along the axis of the TRUS transducer and three rotations (Figure 6.1). The subsystem has a cradle that accommodates any commercially available end-firing TRUS transducer. Because the tracking subsystem has encoders at the joints and in the cradle, motion of the transducer is tracked allowing real-time tracking and recording of the 3D position and orientation of the TRUS transducer. With this arrangement, the biopsy needle location in space (or in the patient) is known as the physician manipulates the TRUS transducer. This mechanical subsystem is composed of two mechanisms: (1) an articulated multijointed *arm* and (2) a TRUS transducer-tracking mechanism, described as follows.

The mechanical tracking subsystem is composed of a spherical linkage assembly so that the axis of the joints converges and constrains the transducer rotation about a common point, which is the RCM (Figure 6.1b). Thus, after the transducer is inserted into the rectum with its tip positioned beneath the prostate against the rectal wall, any forces from rotation of the transducer are minimized on the prostate, reducing the possibility of prostate motion. In addition, deformation of the prostate is also minimized during the biopsy procedure allowing accurate guidance of the biopsy needle to its target based on the registered MR image. All joints on the subsystem have angle-sensing encoders and form the TRUS transducer-tracking mechanism. These encoders allow tracking of each linkage by tracking the angles between the arms, which is transmitted to the computer so that the relative position of the transducer in space can be calculated as it is being manipulated.

6.2.2 3D Transrectal Ultrasound Image Acquisition and Reconstruction

To produce a 3D TRUS image, the end-firing TRUS transducer is mounted in a cradle that is connected to the mechanical-tracking subsystem. The physician can rotate the cradle manually around its longitudinal axis (Figure 6.1b), whereas the 2D TRUS images from the US machine are acquired into the computer through a digital video frame grabber. A rotary encoder is used to sense the angular position, which is also sent to the computer, to provide the known angular position of every 2D TRUS image that has been captured. Typically, we acquire the 2D TRUS images in 0.8° angular steps over a 200° rotation in about 8 to 10 s, which are reconstructed into a 3D image as they are acquired. This approach provides the physician with a 3D image immediately after completion of the rotation (Fenster, 2000b).

6.2.3 3D Transrectal Ultrasound to Magnetic Resonance Registration and Guidance Software

The guidance of the biopsy needle to a target identified in an MR image requires accurate registration of the 3D TRUS and MR images. Because different deformation of the prostate in the TRUS- and MR-imaging procedures may have occurred over time due to different patient positions, bladder filling, rectal wall motion, and the TRUS transducer or MR endorectal coil pressure, the registration method must be nonrigid. Nonrigid 3D MR–TRUS registration is made difficult by the different appearances of the prostate in the MR and 3D TRUS images. Additional difficulty is present in identifying the prostate apex and base in the images from the two modalities.

We have developed two 3D TRUS to MR nonrigid registration methods: (1) *intensity based* and (2) *boundary based*. The intensity based method (Sun et al. 2015) used the modality independent neighborhood descriptor (MIND) metric for 3D TRUS to MR nonrigid registration. The experimental results using 17 patients' MR and 3D TRUS images showed that the method yielded an overall mean target registration error (TRE) of 1.76 mm and a Dice similarity index (DSI) of 80.8% for the apex region, 92.0% for the mid-gland region, and 85.7% for the base region. The DSI for the whole gland was 85.7%. Although this method produced good results, our findings showed that its performance suffered when the TRUS image contained artifacts, such as shadows.

Because of the sensitivity of the intensity based features to 3D TRUS image quality, we developed the *boundary based* registration method using the segmented prostate boundaries in the 3D TRUS and MR images (Qiu et al. 2014a, b, 2015). In this approach, the prostate in the MR and 3D TRUS images are initially aligned rigidly followed by a nonrigid registration step. The rigid alignment procedure involves the user manually identifying six corresponding landmarks in the two 3D images: the end point of peripheral zone at the apex, leftmost, rightmost, topmost, and bottommost points on the largest view of the axial slices, and the urethra at its entrance into the prostate. The prostates' boundaries are then segmented and radially sliced around a rotational axis to generate corresponding prostate slices with segmented boundaries. In addition, equally spaced points on the boundaries are then calculated and used to perform a registration on points from all slices in the two images (Sun, Qiu et al. 2015a) (Figure 6.2). The evaluation of this method using 3D TRUS and MR prostate images of 17 patients showed that the TRE was 3.50 mm for the rigid registration and 2.24 mm for the nonrigid registration (Sun, Qiu et al. 2015b).

The prostate biopsy guidance interface (Figure 6.2) is composed of four windows: the live 2D TRUS video stream from the US machine; a 3D TRUS image with the tracked orientation and position of the transducer during manipulation to allow for context and comparison of the 3D image with the real-time 2D image to determine if the prostate has

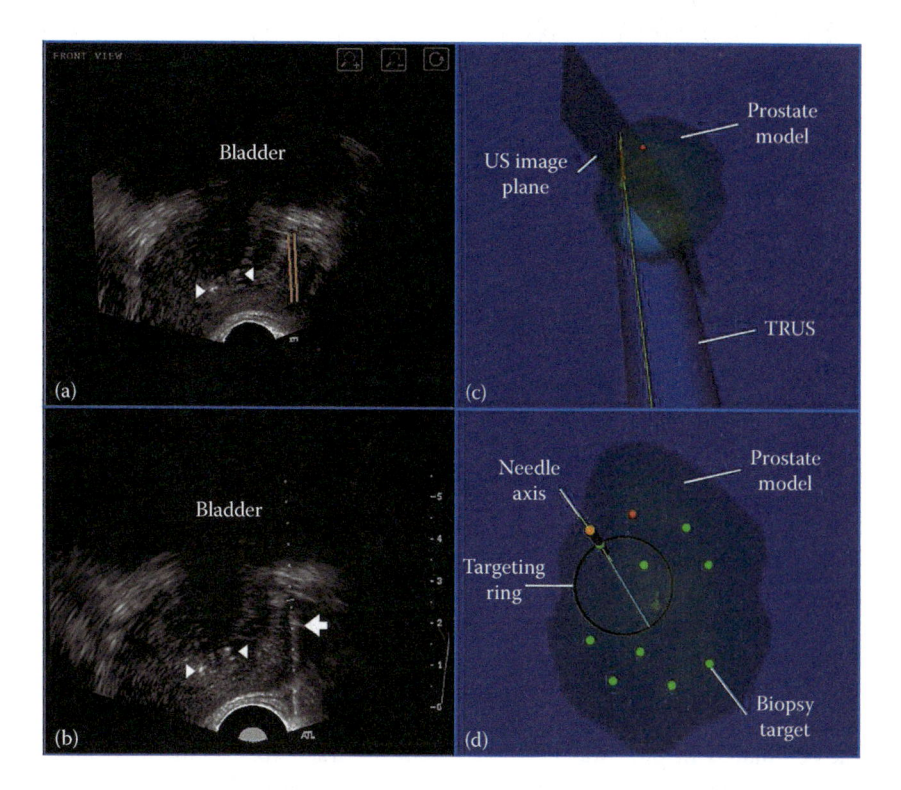

FIGURE 6.2 A view of the four windows in the 3D TRUS-guided prostate biopsy system interface: (a) the 3D TRUS image automatically sliced to match the real-time 2D TRUS transducer orientation, (b) the live 2D TRUS video stream from the US machine, and (c and d) two panels showing the 3D location of the biopsy core within the 3D prostate models. The green dots are targets to be biopsied and the red dot is the next location to be biopsied.

moved or deformed; and two 3D graphical navigation views showing the prostate model and selected targets and the trajectory of the needle if it were fired.

6.2.4 Motion Compensation Using 2D Transrectal Ultrasound to 3D Transrectal Ultrasound Registration

The accurate guidance of the biopsy needle is also dependent on the intraoperative registration to the live 2D TRUS image when prostate motion is present. After nonrigid registration of the 3D TRUS and MR images, a registration can be performed between the 3D TRUS and 2D TRUS images to superimpose the MR-identified contours and targets onto the live 2D image. Unfortunately, prostate motion will cause misalignment of the identified targets and contours, which may lead to an error in targeting the suspicious tissue, and therefore not collecting the appropriate tissue for histological analysis. This prostate motion could be in the form of incremental shifts that result from patient's unease or sudden jumps due to pain or audible snapping of the biopsy gun. Thus, intraoperative motion compensation is required to accurately link prior images and information needed to target and obtain samples of suspicious cancer tissues.

We have developed two real-time 3D TRUS to 2D TRUS intensity based rigid registration methods: (1) a single, intermittent user-initiated registration and (2) an automatic, continuous registration (Gillies et al. 2017). Both methods used the normalized cross-correlation similarity metric and were optimized using the Powell's method, which sequentially searches one DOF at a time to find the optimum of the search space. The 2D and 3D TRUS images were loaded into a graphics processing unit (GPU) to perform parallelization of the cross-correlation metric computation, which decreases the computation time needed for the overall registration. In addition, these images were downsampled by a factor of four and cropped to a bounding rectangular region of interest at the approximate prostate boundaries to only pass signal data within the average prostate volume, which reduced the input information and computation cost of the registration algorithm.

Encoders on the mechanical tracking subsystem were used to initialize the pose of the 3D TRUS image and used as an input for the registration algorithm. When performing a user-initiated registration on a clinical dataset, experimental results from 14 patient's 3D TRUS and 2D TRUS image pairs showed that the method yielded an overall mean TRE of 1.40 mm and a median computation time of 55 ms (Gillies et al. 2017). These findings were very promising, but to improve clinical workflow, an automatic and continuous method was required to run in the background without any user interaction.

The continuous method was developed from the framework of the user-initiated method by adding an additional process to the registration algorithm. After completing an initial user-initiated registration, the resulting registration-correction matrix was now considered as the initialization pose for the subsequent image frame without relying on the tracking subsystem. By performing this constant iterative approach to registration, the correction distance required between frames would likely be minor when using an US system with a frame rate of approximately 30 Hz for one focal point and a depth of 6 cm. In lieu of a clinical dataset, the motion compensation range of this method was tested and compared to the user-initiated method on an agar-based tissue-mimicking phantom.

Both methods were performed to correct for maximum translation (inplane and out-of-plane) and rotation (roll around the transducer long axis) displacements up to 12 mm and 15°, respectively, after mounting the phantom on a micrometer-driven movable stage to provide ground truth displacements. The continuous method performed registration significantly faster ($p < 0.05$) than the user-initiated method with observed computation times of 35 ± 8 ms, 43 ± 16 ms, and 27 ± 5 ms for inplane, out-of-plane, and roll motions, respectively (Gillies et al. 2017). This was in contrast to the user-initiated computation times of 108 ± 38 ms, 60 ± 23 ms, and 89 ± 27 ms (Gillies et al. 2017). Although a reduction in computation times were observed, both methods performed with similar registration errors. Errors for the continuous method were computed as 0.2 ± 0.3 mm, 0.7 ± 0.4 mm, and $0.8 \pm 1.0°$ for inplane, out-of-plane, and roll motions, respectively, and 0.4 ± 0.3 mm, 0.2 ± 0.4 mm, and $0.8 \pm 0.5°$ for the user-initiated method (Gillies et al. 2017).

6.3 3D Ultrasound-Guided Focal Liver Tumor Ablation

Hepatocellular carcinoma (HCC) is the fifth most commonly diagnosed cancer in men and seventh in women, and the third most frequent cause of cancer-related deaths worldwide (Jemal et al. 2011). Incidence is high in Asia and Africa due to endemic viral hepatitis B and C, and it is also increasing in developed nations (El-Serag et al. 2007; Altekruse et al. 2009; Jemal et al. 2010). Moreover, the liver is the second most common site of metastatic cancer from other organs. Although resection and liver transplant for HCC and resection for metastases are standard therapies, only 15% of patients are eligible (El-Serag et al. 2008), and the rest are usually offered chemotherapy and radiotherapy. Because these therapies have limited success rates, some patients are offered image-guided percutaneous focal ablation with the intent to cure or extend life.

The current standard image-guided percutaneous focal liver tumor ablation uses CT images for the diagnosis and planning, and 2D US for intraoperative guidance of the radiofrequency (RF) and microwave (MW) thermal therapy ablation applicators(s) into the tumor. These techniques suffer from local recurrence rates primarily due to (1) requirement of physicians to mentally integrate many 2D US images to form an impression of the tumor size and shape, nearby vasculature, and location of applicators and their tips, leading to variability in planning, guidance, and verification of the applicators' proper location; (2) liver motion caused by breathing that reduces targeting accuracy; and (3) the use of 2D US to measure tumor shape and volume needed for treatment planning resulting in variability and inaccuracy. Some recent approaches have used registered 2D US with preoperative CT images using electromagnetic tracking to guide the procedure. However, this approach is limited because of the requirement for a controlled surgical environment without electromagnetic interference and ferrous materials, which is difficult to implement in countries where HCC is highly prevalent.

To overcome these limitations, we have developed a 3D US-based mechatronic needle-guidance system that can register 3D US with CT images, combined with guidance software, therapy applicator tracking, and motion compensation for an accurate and precise method for the complete ablation of liver tumors.

6.3.1 3D Ultrasound-Guided System

Our liver focal ablation guidance system we developed has the following key elements: (1) a mechanical 3D US scanning mechanism that is compatible with any US system manufacturer's US transducer. The mechanism has a motor that translates the transducer over 5 cm while tilting over 60° generating an image of a large volume; (2) a software application that acquires conventional 2D US images from the US machine and reconstructs these images into a 3D US image as the images are acquired; (3) a mechanical guidance subsystem, rather than electromagnetic, making the system less costly and without background metal sensitivity. This approach makes the system more accessible to a wider range of therapy sites, especially in developing countries. The guidance subsystem is a multijoint *arm* composed of three counterbalanced links with encoders in the joints. The third link holds the 3D scanner subsystem allowing manipulation of the scanner over the patients' abdomen to image the liver; and (4) 3D US viewing software, which allows manipulation of the 3D US image and slicing in any plane, including oblique to allow viewing of the complete tumor and the nearby vasculature.

6.3.2 Experimental Results of the Liver-Ablation System

We have evaluated the accuracy of the mechanical scanner and 3D reconstruction with a 3D grid of strings and 20 cm³ spherical agar objects. Our results of these tests have shown that the system is accurate with linear errors of about 0.3 mm over a distance of 10 mm, and volumetric errors of about 3%. We have also evaluated the system by imaging patients with two RF applicators inserted into their livers and imaged with both CT and 3D US (Figure 6.3). We compared the angles between the applicators and their tips are identified in the two imaging modalities, and showed that the mean angular difference between CT- and 3D US-based measurements was 4.05° ± 3.5°, and the mean applicator tip difference was 2.1 ± 1.3 mm.

Our system promises to provide better treatment options for nonsurgical candidates, and allows treatment of difficult and otherwise untreatable lesions. As our approach makes use of existing US systems and is based on mechatronic guidance, it will be accessible to a wider range of clinics in Canada and developing countries.

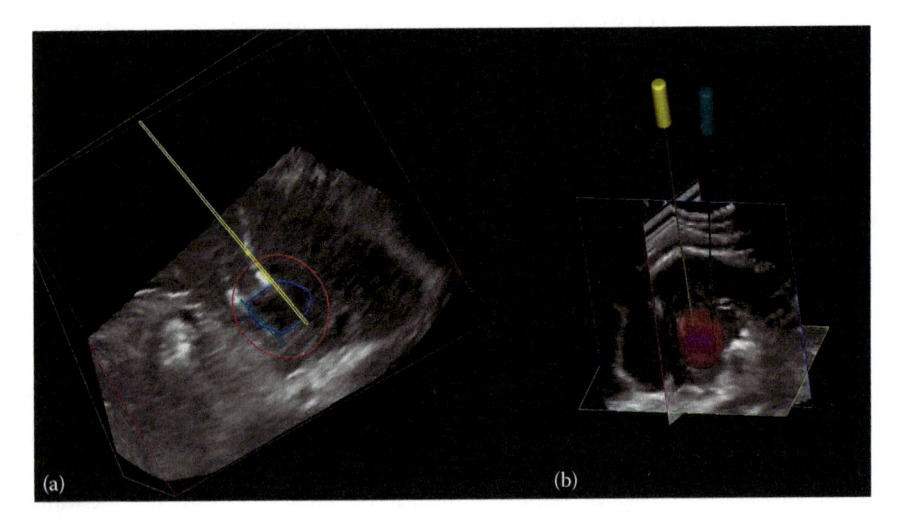

FIGURE 6.3   3D ultrasound image of a patient's liver with RF applicators inserted: (a) shows one applicator in the liver with the segmented tumor in blue and the manufacturer's ablation zone in red and (b) shows a cross-sliced 3D US image with two applicators rendered in yellow and green, and the thermal ablation zone outlined in red.

6.4 3D Ultrasound-Guided Interstitial Gynecological Brachytherapy

Gynecological malignancies are among the most prevalent cancers in women, where cervical cancer alone is the fourth most common cancer and the fourth leading cause of cancer deaths in women worldwide, whereas this rises to the second highest incidence and third highest mortality across women in developing countries (American Cancer Society, 2015). The standard-of-care for the treatment of these gynecological cancers often includes a combination of external-beam radiation and brachytherapy, which allows a higher radiation dose to be delivered to the tumor in comparison to the surrounding tissues. In some cases of cervical cancer, vaginal cancer, and recurrent gynecological cancers a high-dose-rate (HDR) interstitial-brachytherapy approach may be indicated (Viswanathan et al. 2012). In this procedure, hollow needles are inserted into or near the tumor through a template on the patient's perineum and a high-activity radioactive source is temporarily placed at planned positions within these needles to deliver the treatment. Precise needle placement is required for the optimal treatment and avoidance of nearby organs-at-risk, particularly the bladder and rectum (Viswanathan et al. 2012).

Needle placement is commonly verified on a postinsertion CT scan acquired once all needles have been placed; however, misplaced needles identified at this time must either be reinserted or will not be useful for treatment delivery (Viswanathan et al. 2011). Despite the importance of the correct needle placement, there is currently no standard imaging modality for intraoperative visualization as the needles are inserted. In some cases, a 2D transabdominal US may be used to visualize anteriorly placed needles and avoid the bladder (Erickson et al. 1995) and 2D TRUS (Stock et al. 1997; Weitmann et al. 2006; Sharma et al. 2010) is sometimes used to assess needle positions intraoperatively. Typically, a biplane probe is used in a 2D TRUS approach to needle visualization where the transverse plane is used to localize the needle and then the operator switches to longitudinal, side-fire plane to assess needle depth. This approach is limited, as the clinician must mentally transform the 2D views to interpret the position of the needle and relevant structures in 3D space, a process that is inefficient and leaves the localization prone to inaccuracy and variability. Some techniques have been proposed using intraoperative CT (Lee et al. 2013) or MR (Kapur et al. 2012; Viswanathan et al. 2006, 2013) methods to provide 3D image guidance; however, these techniques are restricted by the need for highly specialized equipment and procedure suites, limiting accessibility to only a few institutions (Viswanathan et al. 2011).

To address these limitations, we have developed a 3D US system with transrectal and transvaginal imaging modes to provide visualization of needles intra-operatively during HDR interstitial gynecologic brachytherapy using a clinical 2D side-fire TRUS transducer and compatible with the current perineal needle templates. Our aim is to incorporate this system with guidance software to provide needle tracking and integration with preoperative imaging to provide a more tailored and precise method for gynecologic brachytherapy needle insertions.

6.4.1 3D Ultrasound System

To obtain a 3D US image, the conventional 2D side-fire TRUS transducer is held in a cradle, which is connected to a motorized mover linked to a desktop computer. The clinical US machine is also connected to the computer through a digital video frame grabber, which is used to transmit the 2D US frames. These 2D US images are reconstructed into the 3D US image, which is available immediately after the completion of the scan. In the TRUS scanning mode, the rotational mover turns the transducer cradle 170° along its longitudinal axis in 12 s, acquiring 300 2D frames with 0.57° angular intervals between frames, following a tilt-scanning approach (Fenster et al. 2001) to create a fan-shaped image. Similarly, in the trans-vaginal US (TVUS) scanning mode, the cradle is rotated 360° in 24 s, acquiring 536 2D US frames (angular intervals of 0.67°) to create a ring-shaped image.

6.4.2 Experimental Results of the 3D Ultrasound System

The geometric accuracy of the 3D image reconstruction was validated for both scanning modes using a 3D 10 mm grid phantom. The distance errors associated with the 3D TRUS mode were within 0.3 mm of the expected distance, whereas the distance errors associated with the 3D TVUS mode were within 0.15 mm of expected distance. The volumetric accuracy of the 3D TRUS mode was validated using an agar model of a tumor with a volume of 46.57 cm^3, segmented from a patient MR image, whereas the volumetric accuracy of the 3D TVUS mode was validated using an agar phantom containing a 22.45 cm^3 sphere. The mean volumetric error in the 3D TRUS mode was 0.2% of the expected value. For the spherical phantom, the mean volume measured with the 3D TVUS mode was 2.65% of the nominal volume with a surface area to volume ratio within 0.99% of the expected value, indicating that the shape appeared spherical in the 3D image.

After evaluating the geometric accuracy of the 3D US system, an agar phantom mimicking relevant features of the female pelvis was designed to assess the feasibility of visualizing interstitial brachytherapy needles. This phantom included a vaginal canal and rectal canal to allow for the 3D US scans to be acquired, and a model uterus and tumor for anatomical detail, and included a template through which needles were inserted in a typical configuration. An MR image was also acquired of the needle tracks to allow for comparison of the needle trajectories and tips between imaging modalities. In a 3D TRUS, the mean difference in the needle trajectory was 0.94° ± 0.89° and mean difference in the needle tip position was 1.54 ± 0.71 mm between the modalities (Rodgers et al. 2017). An example of the 3D TVUS image of the pelvic phantom is shown in Figure 6.4. The mean needle trajectory difference was 1.07° ± 0.75° and the mean tip difference was 2.11 ± 0.73 mm in comparison to the MR image.

The 3D TRUS mode of the system was also evaluated in a proof-of-concept study of five patients and the needle trajectory and tip position was compared between the intraoperative 3D TRUS (Figure 6.5) and postinsertion CT image. A total of 73 needles were placed,

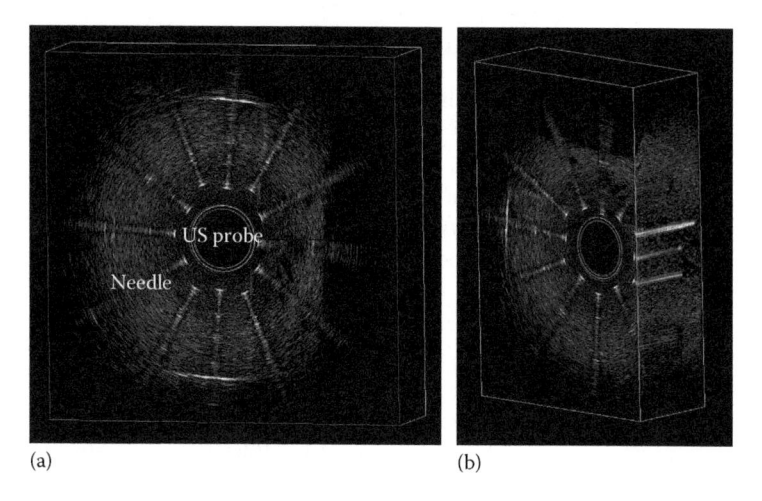

(a)   (b)

FIGURE 6.4 Phantom 3D TVUS image with features indicated. (a) Slice through the 3D TVUS image orthogonal to the US probe showing the needles and a central black region where the US probe was inserted through the hollow cylinder, and the brachytherapy needles surrounding the probe and (b) slice parallel to the needles showing the extent of three needles.

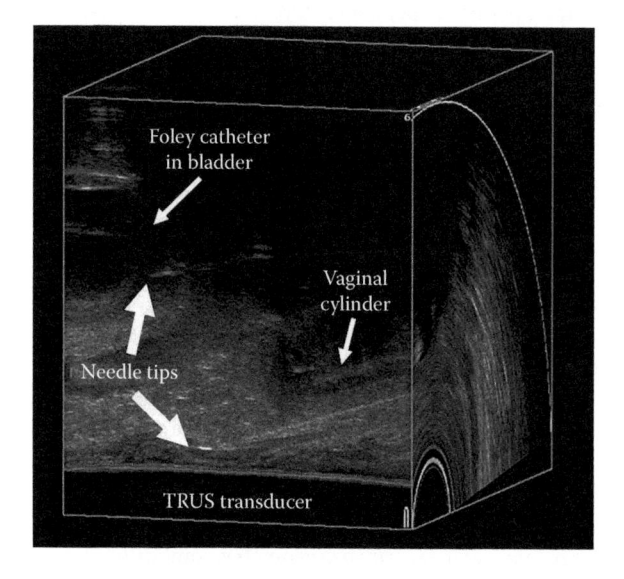

FIGURE 6.5 Patient 3D TRUS image with key features, including two visible needles indicated.

where 88% of the needles were visible on the 3D TRUS images and 79% of the needle tips were identifiable. The mean trajectory difference between the two modalities was $3.04° \pm 1.63°$ and the mean needle tip difference was 3.82 ± 1.86 mm. The 3D TVUS mode of the system will also be evaluated in a similar *in vivo* proof-of-concept study in the future.

6.5 3D Ultrasound-Guided Breast Brachytherapy

255,000 cases of breast cancer are diagnosed annually in Canada (Canadian Cancer Society's Advisory Committee on Cancer Statistics 2015) and the United States (DeSantis et al. 2014), the majority of which are diagnosed at an early stage (Njeh et al. 2010) (Canadian Partnership Against Cancer 2015). Among early stage breast cancers, the standard of care is breast-conserving therapy (BCT) (Goyal et al. 2015), in which the tumour is surgically removed (lumpectomy) before an adjuvant whole breast irradiation (WBI). BCT is an alternative to mastectomy, in which the entirety of the affected breast is removed, and has been shown to provide equivalent rates of cancer recurrence (Fisher et al. 2002). However, WBI is lengthy, requiring 3.5–7 weeks of therapy (Cox and Swanson 2013), and its length is widely thought to cause some women to decline BCT (Wazer et al. 2009), opting for mastectomy or lumpectomy alone instead.

As a result, accelerated partial breast irradiation (APBI), the accelerated delivery of radiotherapy to only the area surrounding the original tumour site, has become an active area of research. One method of APBI known as the permanent breast seed implantation (PBSI), reduces radiation treatment time to a single visit by permanently implanting Pd-103 radioactive seeds, with five-year follow-up of 134 PBSI patients showing recurrence rates equivalent to WBI (Pignol et al. 2015). However, PBSI suffers from procedural complexity and operator dependence. It requires an average of 13–21 seed-bearing needles, inserted through a template consisting of a rectangular grid of needle holes, to deliver approximately 55–95 seeds (Keller et al. 2012). To address the high-operator dependence of the procedure, additional intraoperative guidance is required.

6.5.1 3D Ultrasound Guidance System: Scanner and Tracking Arm

We have developed an intraoperative guidance system (Figure 6.6) that consists of two subsystems: (1) a 3D US scanner and (2) a template tracking arm. The guidance system tracks the position and orientation of the needle template and visualizes it in a common coordinate system with an intraoperative 3D US image.

The 3D US scanner consists of a motorised assembly connected to a standard 2D US probe. Although the scanner could be modified to accommodate any 2D transducer and US system, experiments were performed with a Philips iU22 and an L12-5 linear transducer. One hundred and fifty 2D images are reconstructed into a 3D volume as they are acquired, producing a 3D image that is available immediately after the 10–18 s scan.

The 3D image is reconstructed in a *hybrid* geometry, combining 50 mm of linear translation with 60° of tilt for a transducer 50 mm wide. The scanner is calibrated to a range of depth settings between 4 and 8 cm, resulting in a 3D volume approximately trapezoidal in shape (Figure 6.7) with a bounding box ranging from 90 mm × 50 mm × 40 mm in size to 130 mm × 50 mm × 80 mm. A thin plate of polymethylpentene (TPX™) plastic is mounted on the scanner between the moving probe and the patient or phantom being imaged (*subject*) with US-coupling gel applied to both sides. TPX is sonolucent, allowing US images to be captured through the plate, whereas the stiff plastic prevents tissue motion during the scan.

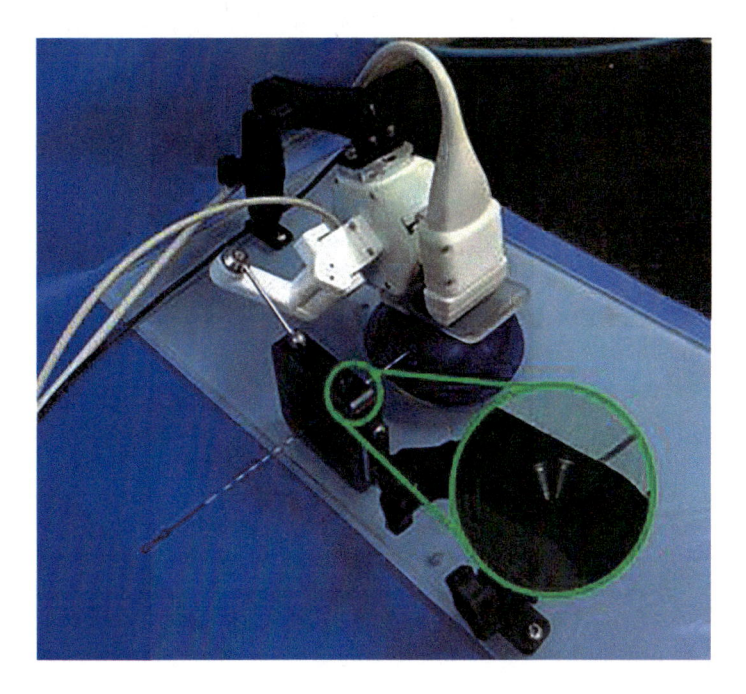

FIGURE 6.6 3D US scanner, encoded arm and trackable template. Breast phantom visible beneath scanner. Magnified window shows one of four fiducial divots on the edge of template (two on top, two on the left edge of the template not visible in image).

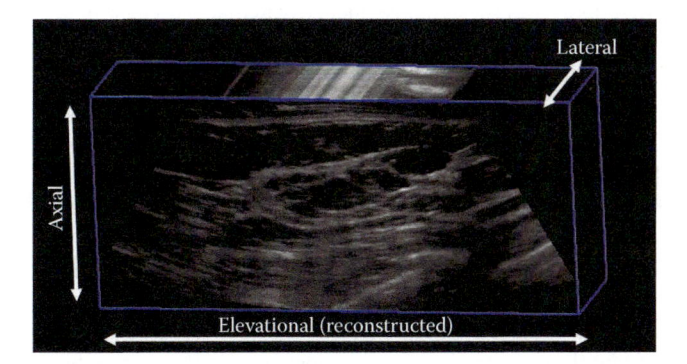

FIGURE 6.7 Visualization of 3D US volume of healthy volunteer taken to assess clinical suitability of the scanner. Visible cross section indicates shape of the volume resulting from *hybrid* scan geometry.

The template-tracking arm is attached to the 3D US scanner and consists of three rigid segments, each rotating passively about a one DOF joint tracked by an absolute encoder. Encoder measurements are used to calculate the position of the spherical tip of the arm relative to its base. A predetermined transform relates the position of the arm with respect to its base to its position with respect to the 3D US image. The arm's spherical tip mates with four fiducial divots on the outside edge of the needle template (magnified window, Figure 6.6), registering the template to the 3D image.

Using the developed system, the proposed workflow is to position the 3D US scanner over the breast using a counterbalanced and adjustable locking arm. The tracking arm is then used to register the template relative to the intraoperative 3D US image. Associated guidance and visualisation software overlays intraoperative 3D images with expected needle trajectories along any plane the user chooses.

6.5.2 Experimental Results of the 3D Ultrasound Guidance System

The 3D US scanner was validated for (1) linear measurement accuracy, (2) volumetric accuracy, and (3) clinical feasibility. Linear measurement accuracy was validated using a 3D grid phantom of strings separated by 10 mm in each of three orthogonal directions and repeated at all depth settings of the 2D US described earlier. Mean measurements for all directions and depth settings were within ±1.4% (0.14 mm) of nominal.

Volumetric accuracy was validated using two agar phantom casts in moulds matching with the patient contours of postlumpectomy surgical cavities, the radiation target for PBSI. Phantoms were scanned and segmented and their volumes compared to water displacement measurements. Mean volumetric measurements were within ±4.1% (± 0.16 cm^3) of water displacement measurements.

A clinical feasibility study was conducted by recruiting a healthy female volunteer whose breasts were imaged using the device by an experienced ultrasonographer. 3D US images (Figure 6.7) were qualitatively assessed by the ultrasonographer as being of clinically useful quality and showed no evidence of artifacts because of probe motion or inadequate coupling.

With the 3D US scanner validated, the template-tracking system was calibrated and its accuracy assessed. Calibration of the tracking arm included registration of the arm with the 3D US image, which was conducted in two parts: (1) registration of the arm to the scanner and (2) registration of the scanner to the 3D US image.

Registration of the arm to the scanner was conducted using a machined calibration jig (Figure 6.8) with 45 fiducial divots. The scanner was securely mounted to the jig in four different positions (translated left/right and forward/back; 3 cm separation), creating 180 known positions, 60 of which were used as calibration points to perform the registration. The registration was also used to determine the encoder measurements corresponding to 0° (*offsets*) for the two encoders that are most distal to the scanner by optimising the offsets to minimise fiducial registration error (FRE). Point measurement accuracy of the arm and its registration to the scanner were evaluated using the remaining 120 positions. The above-mentioned experiments were conducted after having calibrated the arm encoders *in situ*. The resulting registration yielded a TRE of 0.65 mm (FRE 0.26 mm).

With the arm registered to the scanner, the scanner was registered to the 3D image using a second machined test jig with the 3D US scanner mounted to the outside and a string phantom mounted to the inside (Figure 6.9). The string phantom creates 12 string intersections at known locations that are relative to the scanner to facilitate registration.

FIGURE 6.8 Guidance system mounted to calibration jig for the localizing arm. Jig has 45 divots (9 rows × 5 columns) matching with the geometry of fiducial points on the modified needle template. Jig used to calibrate arm, register arm to the scanner.

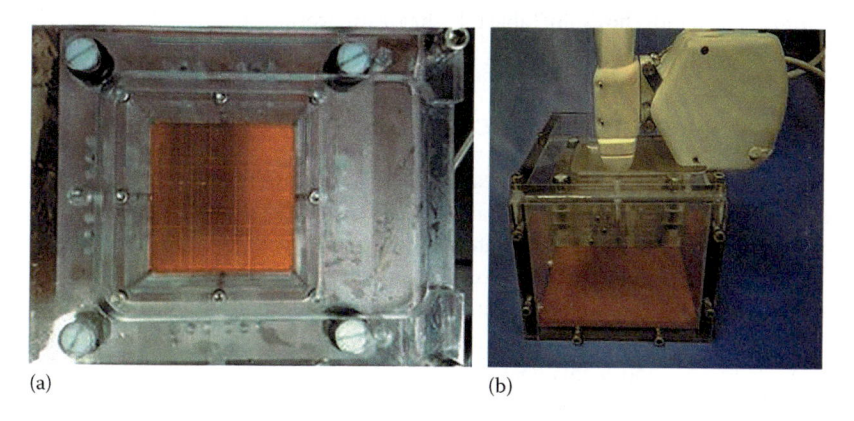

(a)   (b)

FIGURE 6.9 Scanner to 3D image registration jig: (a) view of intersecting strings from above and (b) scanner mounted to jig over-top of strings. When used, container is filled with isopropyl alcohol solution (IAS) (speed of sound = 1540 m/s).

TABLE 6.1   Needle Tip Error between Tracking
and Imaging for Each Insertion Depth

Needle Insertion Depth (cm)	Magnitude of Needle Tip Error (mean ± SD) (mm)
11	2.22 ± 0.41
13	1.61 ± 0.86
15	3.20 ± 0.79
17	2.80 ± 0.60
ALL	**2.46 ± 0.88**

Note: $N = 5$ needles per insertion depth, 20 total.

Six of the intersections were used as registration fiducials, whereas the other six were used as targets to assess registration accuracy. When in use, the jig is filled with 7.25% isopropyl alcohol solution (IAS), matching the nominal speed of sound in tissues (1540 m/s) (Thouvenot et al. 2016). Results of the registration showed a TRE of 0.65 mm and an FRE 0.26 mm.

With the 3D US images validated for reconstruction accuracy and registered with the tracking arm, the imaging and tracking subsystems could be integrated to evaluate the guidance system's tracking error. A mock brachytherapy needle was inserted at a known distance through a tracked needle template and its expected tip position from tracking was compared to its observed tip position under imaging. Needles were inserted into IAS to provide speed of sound matching without causing needle deflection. Needles were inserted at five positions on the needle template forming the corners and center of a rectangle 50 mm wide and 30 mm tall at each of four insertion depths (11, 13, 15, and 17 cm). The template positions and needle-insertion depths were chosen based on treatment plans from patients treated at the Centre for the Southern Interior in Kelowna, Canada. The mean needle tip error across all insertion depths was 2.46 mm with needle tip error at each depth shown in Table 6.1. The future work includes conducting a phantom and then clinical procedures with the integrated system.

6.6 Discussion and Conclusion

We have demonstrated the technical feasibility of 3D US-based brachytherapy and focal ablation procedures. The experimental results indicate that, with mechatronic assistance, the needle can be guided accurately and consistently to target points in the 3D US image as specified by a preplan. We expect that with the introduction of mechatronic system and 3D US-based tools for automatic needle detection and verification of the needles' locations, an effective interventional procedure will be possible. Although most of the tools to accomplish this goal have been developed, additional features are still required for a complete intraoperative system. However, based on the work to-date, we can draw the following conclusions.

In registering the coordinate systems of the mechatronic system with the 3D US image coordinate system, fiducial points in the 3D US image must be determined. Poorer resolution in the 3D scanning direction may result in errors in locating these fiducial points.

These errors will be greater for the points farther away from the US transducer as the image resolution degrades with distance. Calibrating the robot coordinate system and integrating it with the 3D US image requires that fiducial points related to the robot be identified.

Needle-segmentation methods often assume that the needles are straight. However, large implantation errors are primarily caused by needle deflection from the planned trajectory because of the bevel at the needle tip. Thus, the segmentation methods should make use of this fact and use a method that does not assume a straight needle after it has been implanted.

Visualization of the needle trajectory and tip is critical for US-guided brachytherapy and focal ablation procedures. Several factors hamper the accuracy and consistency of needle trajectory and tip visualization in the US images. The main factor for loss of needle US image signal is the alignment of the needle with the US transducer. The signal is greatest when the needle is perpendicular to the US beam, and is reduced as the angle is increased from perpendicular. This effect will not be problematic in the gynecologic and breast brachytherapy procedures, but will be a significant problem in the liver system. Alignment of the needle can be solved with guides and tracking devices (e.g., electromagnetic, optical, acoustic, and mechanical tracking); however, they add complexity and cost to the procedure making them less desirable. Doppler approaches have been used to identify the movement of the needle tip; however, opportunities are still available to develop robust and cost-effective solution to make needles more visible or localizable when inserted into tissues at various angles.

Our patient studies have shown that it is possible to minimize the effects of prostate motion using an image-based approach. Clinical studies are still needed to verify this approach. In addition, extension to other organs (e.g., liver, breast) is still needed.

Acknowledgments

The authors acknowledge the financial support from the Ontario Institute for Cancer Research, Canadian Institutes of Health Research (CIHR), and the Ontario Research Fund (ORF) program. J. Rodgers and D. Gillies acknowledge the support from the Natural Sciences and Engineering Research Council of Canada (NSERC) Canada Graduate Scholarship (CGS).

References

Abbas, F., and Scardino, P. T. 1997. The natural history of clinical prostate carcinoma. *Cancer,* 80(5):827–833.

Altekruse, S., McGlynn, K. A., and Reichman, M. E. 2009. Hepatocellular carcinoma incidence, mortality, and survival trends in the United States from 1975 to 2005. *Journal of Clinical Oncology,* 27:1485–1491.

American Cancer Society. 2015. *Global Cancer Facts & Figures* (3rd ed.). Atlanta, GA: American Cancer Society.

Bax, J., Cool, D. W., Gardi, L., Knight, K., Smith, D., Montreuil, J., Sherebrin, S., Romagnoli, C., and Fenster, A. 2008. Mechanically assisted 3D ultrasound guided prostate biopsy system. *Medical Physics,* 35(12): 5397–5410.

Canadian Cancer Society's Advisory Committee on Cancer Statistics. 2015. Canadian Cancer Statistics 2015. Toronto. https://www.cancer.ca/~/media/cancer.ca/CW/ cancer information/cancer 101/Canadian cancer statistics/Canadian-Cancer-Statistics-2015-EN.pdf.

Canadian Partnership Against Cancer. 2015. Cancer Stage in Performance Measurement: A First Look—A System Performance Spotlight Report. Toronto. http://www.cancerview.ca/idc/groups/public/documents/webcontent/cancer_stage_in_performance_measurement.pdf.

Cox, J. A., and Swanson, T. A. 2013. Current modalities of accelerated partial breast irradiation. *Nature Reviews Clinical Oncology,* 10 (10): 344–356.

DeSantis, C., Jiemin, M., Leah, B., and Ahmedin, J. 2014. Breast cancer statistics, 2013. *CA: A Cancer Journal for Clinicians,* 64 (1): 52–62.

Ding, M., Cardinal, H. N., and Fenster, A. 2003. Automatic needle segmentation in 3D ultrasound images using two orthogonal 2D image projections. *Medical Physics,* 30(2): 222–234.

Draper, K., Blake, C., Gowman, L., Downey, D. B., and Fenster A. 2000, An algorithm for automatic needle localization in ultrasound-guided breast biopsies. *Medical Physics,* 27(8): 1971–1979.

Downey, D. B., and Fenster, A. 1998. Three-dimensional ultrasound: A maturing technology. *Ultrasound Quarterly,* 14, 25–40.

El-Serag, H. B., Lau, M., Eschbach, K., Davila, J., and Goodwin, J. 2007. Epidemiology of hepatocellular carcinoma in Hispanics in the United States. *Archives of Internal Medicine,* 167:1983–1989.

El-Serag, H. B., Marrero, J. A. Rudolph, L., and Reddy, K. R. 2008. Diagnosis and treatment of hepatocellular carcinoma. *Gastroenterology,* 134:1752–1763.

Erickson, B. A., Foley, W. D., Gillin, M., Albano, K., and Wilson, J. F. 1995. Ultrasound-guided transperineal interstitial implantation of gynecologic malignancies: Description of the technique. *Endocurietherapy/Hyperthermia Oncology,* 11(2), 107–113.

Fenster, A., and Downey, D. B. 2000a. Three-dimensional ultrasound imaging. In: Beutel, J., Kundel, H. and Van Metter, R. (Eds.) *Handbook of Medical Imaging, Volume 1, Physics and Psychophysics.* Bellingham, WA: SPIE Press.

Fenster, A., and Downey, D. B. 2000b. Three-dimensional ultrasound imaging. *Annual Reviews of Biomedical Engineering,* 2: 457–475.

Fenster, A., Downey, D. B., and Cardinal, H. N. 2001. Three-dimensional ultrasound imaging. *Physics in Medicine and Biology,* 46, R67–R99.

Fisher, B., Anderson, S., Bryant, J., Margolese, R. G., Deutsch, M., Fisher, E. R., Jeong, J. H., and Wolmark, N. 2002. Twenty-year follow-up of a randomized trial comparing total mastectomy, lumpectomy, and lumpectomy plus irradiation for the treatment of invasive breast cancer. *The New England Journal of Medicine,* 347 (16): 1233–1241.

Gillies, D.J., Gardi, L., Zhao, R., Fenster, A. 2017. Registration Between 2D and 3D Ultrasound Images Real-time for Image-guided Prostate Interventions. *Med Phys,* 44(9): 4708–4723.

Goyal, S., Sheenu, C., Bruce, G. H., and Kitaw, D. 2015. Effect of travel distance and time to radiotherapy on likelihood of receiving mastectomy. *Annals of Surgical Oncology*, 22 (4): 1095–1101.

Hrinivich, W. T., Hoover, D. A., Surry, K., Edirisinghe, C., Montreuil, J., D'Souza, D., Fenster, A., and Wong, E. 2016. Three-dimensional transrectal ultrasound guided high-dose-rate prostate brachytherapy: A comparison of needle segmentation accuracy with two-dimensional image guidance. *Brachytherapy*, 15(2): 231–239.

Jemal, A., Siegel, R., Xu, J., and Ward, E., 2010. Global cancer statistics. *CA: A Cancer Journal for Clinicians*, 60:277–300.

Jemal, A., Bray, F., Center, M. M., Ferlay, J., Ward, E., and Forman, D. 2011. Global cancer statistics. *CA: A Cancer Journal for Clinicians*, 61:69–90.

Kapur, T., Egger, J., Damato, A., Schmidt, E. J., and Viswanathan, A. N. 2012. 3T MR-guided brachytherapy for gynecologic malignancies. *Magnetic Resonance Imaging*, 30(9), 1279–1290.

Keller, B. M., Ananth, R., Raxa, S., and Pignol, J.-P. 2012. Permanent breast seed implant dosimetry quality assurance. *Journal of Radiation Oncology* Biology* Physics*, 83 (1): 84–92.

Lee, L. J., Damato, A. L., and Viswanathan, A. N. 2013. Clinical outcomes of high-dose-rate interstitial gynecologic brachytherapy using real-time CT guidance. *Brachytherapy*, 12(4), 303–310.

Manenti, G., Carlani, M., Mancino, S., Colangelo, V., Di Roma, M., Squillaci, E., and Simonetti, G. 2007. Diffusion tensor magnetic resonance imaging of prostate cancer. *Investigative Radiology*, 42(6):412–419.

Natarajan, S., Marks, L. S., Margolis, D. J., Huang, J., Macairan, M. L., Lieu, P., and Fenster, A. 2011. Clinical application of a 3D ultrasound-guided prostate biopsy system. *Urologic Oncology: Seminars and Original Investigation*, 29(3): 334–342.

Nelson, T. R., and Pretorius, D. H. 1998. Three-dimensional ultrasound imaging. *Ultrasound in Medicine & Biology*, 24, 1243–1270.

Nelson, T. R., Downey, D. B., Pretorius, D. H., and Fenster, A. 1999. *Three-Dimensional Ultrasound*, Philadelphia PA, Lippincott-Raven.

Njeh, C. F., Mark W. S., and Christian M. L. 2010. Accelerated partial breast irradiation (APBI): A review of available techniques. *Radiation Oncology*, 5 (1): 90.

Qiu, W., Yuchi, M., Ding, M., Tessier, D., and Fenster, A. 2013. Needle segmentation using 3D Hough transform in 3D TRUS guided prostate transperineal therapy. *Medical Physics*, 40(4): 042902.

Qiu, W., Yuan, J., Ukwatta, E., and Fenster, A. 2015. Rotationally resliced 3D prostate TRUS segmentation using convex optimization with shape priors. *Medical Physics*, 42(2): 877.

Qiu, W., Rajchl, M., Guo, F., Sun, Y., Ukwatta, E., Fenster, A., and Yuan, J. 2014a. 3D prostate TRUS segmentation using globally optimized volume-preserving prior. *International Conference on Medical Image Computing and Computer-Assisted Intervention*, 17(Pt 1): 796–803.

Qiu, W., Yuan, J., Ukwatta, E., Sun, Y., Rajchl, M., and Fenster, A. 2014b. Dual optimization based prostate zonal segmentation in 3D MR images. *Medical Image Analysis*, 18(4): 660–673.

Pignol, J.-P., Caudrelier J.-M., Crook, J., McCann, C., Truong, P., and Verkooijen, H. A. 2015. Report on the clinical outcomes of permanent breast seed implant for early-stage breast cancers. *International Journal of Radiation Oncology*Biology*Physics*, 93 (3): 614–621.

Rabbani, F., Stroumbakis, N., Kava, B. R., Cookson, M. S., and Fair, W. R. 1998. Incidence and clinical significance of false-negative sextant prostate biopsies. *The Journal of Urology*, 159(4): 1247–1250.

Rodgers, J. R., Surry, K., Leung, E., D'Souza, D., and Fenster, A. 2017. Toward a 3D transrectal ultrasound system for verification of needle placement during high-dose-rate interstitial gynecologic brachytherapy. *Medical Physics*, 44(5): 1899–1911.

Sharma, D. N., Rath, G. K., Thulkar, S., Kumar, S., Subramani, V. and Julka, P. K. 2010. Use of transrectal ultrasound for high dose rate interstitial brachytherapy for patients of carcinoma of uterine cervix. *Journal of Gynecologic Oncology*, 21(1): 12–17.

Silverberg, E., Boring, C. C., and Squires, T. S. 1990. Cancer statistics. *CA: A Cancer Journal for Clinicians*, 40:9–26.

Stewart, B. W., and Wild, C. 2014. World cancer report 2014. *Health*.

Stock, R. G., Chan, K., Terk, M., Dewyngaert, J. K., Stone, N. N., and Dottino, P. 1997. A new technique for performing Syed-Neblett template interstitial implants for gynecologic malignancies using transrectal-ultrasound guidance. *International Journal of Radiation Oncology Biology Physics*, 37(4), 819–825.

Sun, Y., Yuan, J., Qiu, W., Rajchl, M., Romagnoli, C., and Fenster, A. 2015. Three-dimensional nonrigid MR-TRUS registration using dual optimization. *IEEE Transactions on Medical Imaging*, 34(5): 1085–1095.

Sun, Y., Qiu, W., Yuan, J., Romagnoli, C., and Fenster. A. 2015. Three-dimensional non-rigid landmark-based magnetic resonance to transrectal ultrasound registration for image-guided prostate biopsy. *Journal of Medical Imaging*, 2(2).

Thouvenot, A., Poepping, T., Peters, T. M., and Chen, E. C. S. 2016. Characterization of various tissue mimicking materials for medical ultrasound imaging. In D. Kontos, T. G. Flohr, and J. Y. Lo (Eds.), *SPIE Medical Imaging*, 97835E. International Society for Optics and Photonics.

Viswanathan, A. N., Cormack, R., Holloway, C. L., Tanaka, C., O'Farrell, D., Devlin, P. M., and Tempany, C. 2006. Magnetic resonance-guided interstitial therapy for vaginal recurrence of endometrial cancer. *International Journal of Radiation Oncology Biology Physics*, 66(1), 91–99.

Viswanathan, A. N., Erickson, B. E., and Rownd, J. 2011. Image-Based Approaches to Interstitial Brachytherapy. In A. N. Viswanathan, C. Kirisits, B. E. Erickson, and R. Pötter (Eds.), *Gynecologic Radiation Therapy: Novel Approaches to Image-Guidance and Management* (pp. 247–259). Berlin, Germany: Springer.

Viswanathan, A. N., Szymonifka, J., Tempany-Afdhal, C. M., O'Farrell, D. A., and Cormack, R. A. 2013. A prospective trial of real-time magnetic resonance-guided catheter placement in interstitial gynecologic brachytherapy. *Brachytherapy*, 12(3), 240–247.

Viswanathan, A. N., Thomadsen, B., and American Brachytherapy Society Cervical Cancer Recommendations Committee. 2012. American Brachytherapy Society consensus guidelines for locally advanced carcinoma of the cervix. Part I: General principles. *Brachytherapy*, 11(1), 33–46.

Wazer, D. E., Arthur, D. W., and Vicini, F. A. (Eds.). 2009. *Accelerated Partial Breast Irradiation: Techniques and Clinical Implementation*. 2nd ed. Berlin, Germany: Springer.

Weitmann, H. D., Knocke, T. H., Waldhäusl, C., and Pötter, R. 2006. Ultrasound-guided interstitial brachytherapy in the treatment of advanced vaginal recurrences from cervical and endometrial carcinoma. *Strahlentherapie Und Onkologie*, 182(2), 86–95.

Index

Note: Page numbers followed by f and t refer to figures and tables respectively.